MANY
WORLDS
IN ONE

MANY WORLDS IN ONE

...

The Search for Other Universes

...

ALEX VILENKIN

HILL AND WANG

A division of Farrar, Straus and Giroux

New York

Hill and Wang
A division of Farrar, Straus and Giroux
19 Union Square West, New York 10003

Library of Congress Cataloging-in-Publication Data

Vilenkin, A. (Alexander)
    Many worlds in one : the search for other universes / Alex Vilenkin.— 1st ed.
      p.   cm.
    Includes bibliographical references and index.
    ISBN-13: 978-0-8090-9523-0 (hardcover : alk. paper)
    ISBN-10: 0-8090-9523-8 (hardcover : alk. paper)
    1. Cosmology.   2. Cosmogony.   3. Cosmography.   I. Title.

QB981.V526 2006
523.1—dc22

                                        2005027057

Designed by Cassandra J. Pappas
Drawings by Delia Schwartz-Perlov and the author

www.fsgbooks.com

1   3   5   7   9   10   8   6   4   2

Alina

To Alina

Contents

Prologue   3

Part I · GENESIS

1 · What Banged, How It Banged, and What Caused   9
It to Bang

2 · The Rise and Fall of Repulsive Gravity   13

3 · Creation and Its Discontents   21

4 · The Modern Story of Genesis   29

5 · The Inflationary Universe   45

6 · Too Good to Be Wrong   54

Part II · ETERNAL INFLATION

7 · The Antigravity Stone   73

8 · Runaway Inflation   77

9 · The Sky Has Spoken   87

10 · Infinite Islands   93

11 · The King Lives!   102

Part III · PRINCIPLE OF MEDIOCRITY

12 · The Cosmological Constant Problem   121

13 · Anthropic Feuds   128

14 · Mediocrity Raised to a Principle   140

15 · A Theory of Everything   152

Part IV · BEFORE THE BEGINNING

16 · Did the Universe Have a Beginning?   169

17 · Creation of Universes from Nothing   178

18 · The End of the World   194

19 · Fire in the Equations   199

Epilogue   207

Notes   209

Acknowledgments   223

Index   225

MANY WORLDS IN ONE

Prologue

The stunning success of the book took everybody by surprise. The author, a quiet, even demure physics professor named Alex Vilenkin, has become an instant celebrity. His talk show engagements have been booked solid six months in advance. He has hired four bodyguards and has moved to an undisclosed location to avoid paparazzi. His sensational bestseller, titled Many Worlds in One, describes a new cosmological theory that says that every possible chain of events, no matter how bizarre or improbable, has actually happened somewhere in the universe—and not only once, but an infinite number of times!

The consequences of the new theory are mind-boggling. If your favorite football team did not win the championship, don't despair: it did win on an infinite number of other earths. In fact, there is an infinity of earths where your team wins every single year! If your discontent goes beyond football and you are completely fed up with how things are, again Vilenkin's book has something to offer. According to the new theory, most places in the universe are nothing like our Earth and are even ruled by different laws of physics.

The most controversial aspect of the book is the claim that each of us has an infinite number of identical clones living on countless earths scattered throughout the universe. Much sleep has been lost over this issue. People feel their unique identities have been stolen. So attendance at psychoanalysts' offices has doubled, and sales of the book have soared. Using his theory, Vilenkin also predicted that on some earths his book would be a phenomenal success. But to be fair, he had to admit there were infinite others where it would be a complete flop . . .

•

We live in the aftermath of a great explosion. This awesome event, called somewhat frivolously "the big bang," occurred some 14 billion years ago. The whole of space erupted in a hot, rapidly expanding fireball of matter and radiation. As it expanded, the fireball cooled down, its glow steadily subsided, and the universe slowly descended into darkness. A billion years passed uneventfully. But gradually, galaxies were pulled together by gravity, and the universe lit up with myriads of stars. Planets revolving about some of the stars became home to intelligent creatures. Some of the creatures became cosmologists and figured out that the universe originated in the big bang.

Compared with historians and detectives, the great advantage cosmologists have is that they can actually see the past. Light from remote galaxies takes billions of years to reach our telescopes on Earth, so we observe the galaxies as they were in their youth, when their light was emitted. Microwave detectors pick up the faint afterglow of the fireball, yielding an image of the universe at a still earlier epoch, prior to the formation of galaxies. We thus see the history of the universe unfolding before us.

This wonderful vision, however, has its bounds. Even though we can trace the history of the cosmos to less than a second after the big bang, the bang itself is still shrouded in mystery. What triggered this enigmatic event? Was it the true beginning of the universe? If not, then what came before? There is also a fundamental limit to how far we can see into space. Our horizon is defined by the maximum distance light could have traveled since the big bang. Sources more distant than the horizon cannot be observed, simply because their light has not yet had time to reach Earth. This leaves us wondering what the rest of the universe is like. Is it more of the same, or could it be that distant parts of the universe differ dramatically from our cosmic neighborhood? Does the universe extend to infinity, or does it close in on itself, like the surface of the Earth?

These are the most basic questions about the universe. But can we ever hope to answer them? If I claim that the universe ends abruptly beyond the horizon, or that it is filled with water and inhabited by intelligent goldfish, how can anyone prove me wrong? Cosmologists, therefore, focus mostly on the observable part of the universe, leaving it to philosophers and theologians to argue about what lies beyond.

But if indeed our quest must end at the horizon, wouldn't that be a great disappointment? We may discover scores of new galaxies and map the entire visible universe, just as we mapped the surface of the Earth. But to what end? Mapping our own galaxy could serve a practical purpose, since we may want to colonize it some time in the future. But galaxies billions of light-years away are not likely prospects for colonization. At least not in the next few billion years. Of course, the appeal of cosmology is not in its practical utility. Our fascination with the cosmos is of the same nature as the feeling that inspired ancient creation myths. It is rooted in the desire to understand the origin and the destiny of the universe, its overall design, and how we humans fit into the general scheme of things.

Cosmologists who do rise to the challenge of the ultimate cosmic questions lose all their advantage over detectives. They can rely only on indirect, circumstantial evidence, using measurements made in the accessible part of the universe to make inferences about the times and places that cannot be observed. This limitation makes it much harder to prove one's case "beyond a reasonable doubt." But because of remarkable recent developments in cosmology, we now have answers to the ultimate cosmic questions that we have some reason to believe.

The worldview that has emerged from the new developments is nothing short of astonishing. To paraphrase Niels Bohr, it may even be crazy enough to be true. That worldview combines, in surprising ways, some seemingly contradictory features: the universe is both infinite and finite, evolving and stationary, eternal and yet with a beginning. The theory also predicts that some remote regions have planets exactly like our Earth, with continents of the same outline and terrain, inhabited by identical creatures, including our clones, some of them holding copies of this book in their hands. This book is about the new worldview, its origins, and its fascinating, bizarre, and at times disturbing implications.

PART I

. . .

GENESIS

1

What Banged, How It Banged,
and What Caused It to Bang

In the context of inflationary cosmology, it is fair to say that the universe is the ultimate free lunch.　　—ALAN GUTH

On a Wednesday afternoon, in the winter of 1980, I was sitting in a fully packed Harvard auditorium, listening to the most fascinating talk I had heard in many years. The speaker was Alan Guth, a young physicist from Stanford, and the topic was a new theory for the origin of the universe. I had not met Guth before, but I had heard of his spectacular rise from obscurity to stardom. Only a month before, he belonged to the nomadic tribe of postdocs—young researchers traveling from one temporary contract to another, in the hope of distinguishing themselves and landing a permanent job at some university. Things were looking bleak for Guth: at age thirty-two he was getting a bit old for the youthful tribe, and the contract offers were beginning to dry out. But then he was blessed with a happy thought that changed everything.

Guth turned out to be a short, bouncy fellow, full of boyish enthusiasm, apparently untarnished by his long wanderings as a postdoc. From the outset, he made it clear that he was not trying to overthrow the big bang theory. There was no need to. The evidence for the big bang was very persuasive, and the theory was in good shape.

The most convincing evidence is the expansion of the universe, discovered by Edwin Hubble in 1929. Hubble found that distant galaxies are moving away from us at very high speeds. If the motion of the galaxies is traced backward in time, they all merge together at some moment in the past, pointing to an explosive beginning of the universe.

Another major piece of evidence in favor of the big bang is the *cosmic background radiation*. Space is filled with microwaves of about the same frequency as we use in microwave ovens. The intensity of this radiation dwindles as the universe expands; hence what we now observe is the faint afterglow of the hot primeval fireball.

Cosmologists used the big bang theory to study how the fireball expanded and cooled, how atomic nuclei formed, and how the grand spirals of galaxies emerged from featureless gas clouds. The results of these studies were in excellent agreement with astronomical observations, so there was little doubt that the theory was on the right track. What it described, however, was only the aftermath of the big bang; the theory said nothing about the bang itself. In Guth's own words, it did not say "what 'banged,' how it 'banged,' or what caused it to 'bang.'"[1]

To compound the mystery, on closer examination the big bang appeared to be a very peculiar kind of explosion. Just imagine a pin balancing on its point. Nudge it slightly in any direction and it will fall. So it is with the big bang. A large universe sprinkled with galaxies, like the one we see around us, is produced only if the power of the primordial blast is fine-tuned with an incredible precision. A tiny deviation from the required power results in a cosmological disaster, such as the fireball collapsing under its own weight or the universe being nearly empty.

The big bang cosmology simply postulated that the fireball had the required properties. The prevailing attitude among physicists was that physics can describe how the universe evolved from a given initial state, but it is beyond physics to explain why the universe happened to start in that particular configuration. Asking questions about the initial state was regarded as "philosophy," which, coming from a physicist, translates as a waste of time. This attitude, however, did not make the big bang any less enigmatic.

Now Guth was telling us that the veil of mystery surrounding the big bang could be lifted. His new theory would uncover the nature of the bang and explain why the initial fireball was so contrived. The seminar room fell suddenly silent. Everybody was intrigued.

The explanation the new theory gave for the big bang was remarkably simple: the universe was blown up by repulsive gravity! The leading role in this theory is played by a hypothetical, superdense material with some highly unusual properties. Its most important characteristic is that it produces a strong repulsive gravitational force. Guth assumed that there was some amount of this material in the early universe. He did not need much: a tiny chunk would be sufficient.

The internal gravitational repulsion would cause the chunk to expand very rapidly. If it were made of normal matter, its density would be diluted as it expanded, but this antigravity stuff behaves completely differently: the second key feature of the strange material is that its density always remains the same, so its total mass is proportional to the volume it occupies. As the chunk grows in size, it also grows in mass, so its repulsive gravity becomes stronger and it expands even faster. A brief period of such accelerated expansion, which Guth called *inflation*, can enlarge a minuscule initial chunk to enormous dimensions, far exceeding the size of the presently observable universe.

The dramatic increase in mass during inflation may at first appear to contradict one of the most fundamental laws of physics, the law of energy conservation. By Einstein's famous relation, $E = mc^2$, energy is proportional to mass. (Here, E is energy, m is mass, and c is the speed of light.) So

Figure 1.1. A chunk of gravitationally repulsive material.

the energy of the inflating chunk must also have grown by a colossal factor, while energy conservation requires that it should remain constant. The paradox disappears if one remembers to include the contribution to the energy due to gravity. It has long been known that gravitational energy is always negative. This fact did not appear very important, but now it suddenly acquired a cosmic significance. As the positive energy of matter grows, it is balanced by the growing negative gravitational energy. The total energy remains constant, as demanded by the conservation law.

In order to provide an ending for the period of inflation, Guth required that the repulsive gravity stuff should be unstable. As it decays, its energy is released to produce a hot fireball of elementary particles. The fireball then continues to expand by inertia, but now it consists of normal matter, its gravity is attractive, and the expansion gradually slows down. The decay of the antigravity material marks the end of inflation and plays the role of the big bang in this theory.

The beauty of the idea was that in a single shot inflation explained why the universe is so big, why it is expanding, and why it was so hot at the beginning. A huge expanding universe was produced from almost nothing. All that was needed was a microscopic chunk of repulsive gravity material. Guth admitted he did not know where the initial chunk came from, but that detail could be worked out later. "It's often said that you cannot get something for nothing," he said, "but the universe may be the ultimate free lunch."

All this assumes, of course, that the repulsive gravity stuff really existed. There was no shortage of it in science fiction novels, where it had been used in all sorts of flying machines, from combat vehicles to antigravity shoes. But could professional physicists seriously consider the possibility that gravity might be repulsive?

They sure could. And the first to do that was none other than Albert Einstein.

2

The Rise and Fall of Repulsive Gravity

"We have conquered gravity!" the Professor shouted, and crashed to the floor.
— J. WILLIAMS and R. ABRASHKIN,
Danny Dunn and the Anti-Gravity Paint

THE FABRIC OF SPACE AND TIME

Einstein created two theories of stunning beauty that forever changed our concepts of space, time, and gravitation. The first of the two, called the special theory of relativity, was published in 1905, when Einstein was twenty-six and by most standards could be regarded a failure. His fierce independence and his casual class attendance did not make him popular among the professors at Zurich Polytechnic, where he got his diploma. When the time came to apply for jobs, all his fellow graduates were appointed assistants at the Polytechnic, while Einstein failed to get any academic position. He thought himself lucky to have a job as a clerk at the patent office in Berne, which he got with the help of a former classmate. On the positive side, the patent office work was not without some interest and left plenty of time for Einstein's research and other intellectual pursuits. He spent evenings with friends, smoking a pipe, reading Spinoza and Plato, and discussing his ideas about physics. He also played string quintets in the unlikely company of a lawyer, a bookbinder, a schoolteacher, and a prison guard. None of them suspected that their second violin had something profound to say about the nature of space and time.

Einstein completed the special theory of relativity in less than six weeks of frenzied work. The theory shows that space and time intervals do not by themselves have absolute meaning, but rather depend on the state of motion of the observer who measures them. If two observers move relative to one another, then each of them will find that the other's clock ticks more slowly than his own. Simultaneity is also relative. Events that are simultaneous for one observer will generally occur at different times for the other. We do not notice these effects in our everyday life because they are completely negligible at ordinary velocities. But if the speed of the two observers relative to each other is close to the speed of light, the discrepancies between their measurements can be very large. There is one thing, though, that all observers will agree upon: light always travels at the same speed, approximately 300,000 kilometers per second.

The speed of light is the absolute speed limit in the universe. As you apply a force to a physical object, the object accelerates. Its velocity grows, and if you keep up the force, the velocity of the object will eventually approach the speed of light. Einstein showed that it would take increasingly large amounts of energy to get closer and closer to the speed of light, so the limit can never be reached.

Perhaps the best-known consequence of special relativity is the equivalence of energy and mass, expressed in Einstein's formula $E = mc^2$. If you heat an object, its thermal (heat) energy grows, so it should weigh more. This may give you the idea to take a cold shower before you step on the scale. But this trick is likely to decrease your weight by no more than a few billionths of a pound. In conventional units, like meters and seconds, the conversion factor c^2 between energy and mass is very large, and it takes a huge amount of energy to noticeably change the mass of a macroscopic body. Physicists often use another system of units, where $c = 1$, so that energy is simply equal to mass and can be measured in kilograms.* I will mostly follow this tradition and make no distinction between energy and mass.

The word "special" in "special relativity" refers to the fact that this theory applies only in special circumstances when the effects of gravity are

---

*For example, time can be measured in years and distance in light-years. (A light-year is the distance traveled by light in a year.) Then the speed of light is $c = 1$.

unimportant. This limitation is removed in Einstein's second theory, the general theory of relativity, which is essentially a theory of gravitation.

•

The general theory of relativity grew out of a simple observation, that the motion of objects under the action of gravity is independent of their mass, shape, or any other properties, as long as all nongravitational forces can be neglected. This was first recognized by Galileo, who forcefully argued the point in his famous *Dialogues*. The accepted view at the time, that of Aristotle, was that heavier objects fall faster. Indeed, a watermelon does fall faster than a feather, but Galileo realized that the difference was due only to air resistance. Legend has it that Galileo dropped rocks of different weight from the Leaning Tower of Pisa, to make sure that they landed at the same time. We do know that he experimented with marbles rolling down an inclined plane and found that the motion was independent of the mass. He also offered a theoretical proof that Aristotle could not be right. Suppose, says Galileo, that a heavy rock falls faster than a light rock. Imagine then tying them together with a very light string. How will this affect the fall of the heavy rock? On the one hand, the slower-moving light rock should make the fall of the heavy rock somewhat slower than it was before. On the other hand, viewed together, the two rocks now constitute one object that is more massive than the heavy rock was initially, and thus the two rocks together should fall faster. This contradiction demonstrates that Aristotle's theory is inconsistent.

Einstein was pondering this peculiar kind of motion, which is completely independent of what is moving. It reminded him of inertial motion: in the absence of forces, an object moves along a straight line at a constant speed, regardless of what it is made of. In effect, the motion of the object in space and time is the property of space and time themselves.

Here Einstein made use of the ideas of his former mathematics professor, Hermann Minkowski. As a student, Einstein did not think much of Minkowski's lectures, while Minkowski remembered Einstein as a "lazy dog" and did not expect him to do anything worthwhile. To Minkowski's credit, he changed his mind quickly after reading Einstein's 1905 paper.

Minkowski realized that the mathematics of special relativity becomes simpler and more elegant if space and time are not regarded as separate, but are united in a single entity called *spacetime*. A point in spacetime is an *event*. It can be specified by four numbers: three for its spatial location and one for its time. Hence, spacetime has four dimensions. If you had all of spacetime in front of you, then you would know all the past, present, and future of the universe. The history of each particle is represented by a line in spacetime, which gives the position of the particle at every moment of time. This line is called the *world line* of the particle. (George Gamow, one of the founders of the big bang cosmology, called his autobiography *My World Line*.)

The uniform motion of particles in the absence of gravity is represented by straight lines in spacetime. But gravity makes particles deviate from this simple motion, so their world lines are no longer straight. This led Einstein to a truly astonishing hypothesis: even deviant particles with curved world lines might still be following the straightest possible paths in spacetime, but the spacetime itself must be curved around massive bodies. Gravity, then, is nothing but the curvature of spacetime!

The distortion of spacetime geometry by a massive body can be illustrated by a heavy object resting on a horizontally stretched rubber sheet (see Figure 2.1). The rubber surface is warped near the object, just as spacetime is warped near a gravitating body. If you try playing billiards on this rubber sheet, you will discover that the billiard balls are deflected on the curved surface, especially when they pass near the heavy mass. This analogy is not perfect—it illustrates only the warping of space, not that of spacetime—but it does capture the essence of the idea.

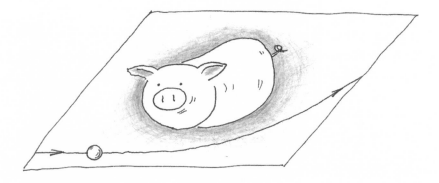

Figure 2.1. A massive body causes space to curve.

It took Einstein more than three years of truly heroic effort to express these ideas in mathematical terms. The equations of the new theory, which he called the general theory of relativity, relate the geometry of spacetime to the matter content of the universe. In the regime of slow motion and not-too-strong gravitational fields, the theory reproduced Newton's law, with the force of gravity being inversely proportional to the square of the distance. There was also a small correction to this law, which was utterly negligible for planetary motion, except in the case of Mercury, the planet closest to the Sun. The effect of the correction was to cause a slow precession, or advance, of Mercury's orbit. Astronomical observations did in fact show a tiny precession, which remained unexplained in Newton's theory, but was in perfect agreement with Einstein's calculation. At this point Einstein was certain that the theory was correct. "I was beside myself with ecstasy for days," he wrote to his friend Paul Ehrenfest.[1]

Perhaps the most remarkable thing about the general theory of relativity is how little factual input it required. The essential fact that Einstein placed at the foundation of the theory—that the motion of objects under the action of gravity is independent of their mass—was known already to Galileo. With this minimal input, he created a theory that reproduced Newton's law in the appropriate limit and explained a deviation from this law. If you think about it, Newton's law is in some sense arbitrary. It states that the gravitational force between two bodies is inversely proportional to the sec-

Figure 2.2. Einstein's equations.

ond power of their distance, but it does not say why. It could equally well be the fourth power or the 2.03rd power. In contrast, Einstein's theory allows no freedom. The picture of gravity as curvature of spacetime inevitably leads to Einstein's equations, and the equations yield the inverse square law. In this sense the general theory of relativity not only describes gravity, it *explains* gravity. So compelling was the logic of the theory and so beautiful its mathematical structure that Einstein felt it simply had to be right. In a letter to a senior colleague, Arnold Sommerfeld, he wrote, "Of the general theory of relativity you will be convinced, once you have studied it. Therefore I am not going to defend it with a single word."[2]

THE GRAVITY OF EMPTY SPACE

With his general theory of relativity now complete, Einstein wasted no time in applying it to the entire universe. He was not interested in trivial details, such as the position of this star or that planet. Rather, he wanted to find a solution of his equations that would describe, in broad brushstrokes, the structure of the universe as a whole.

Little was known at the time about the distribution of matter in the universe, so Einstein had to make some guesses. He made the simplest assumption that, on average, matter is uniformly spread throughout the cosmos. There are, of course, local deviations from uniformity, with the density of stars being a little higher in this place and a little lower in that. What Einstein assumed was that if matter is smoothed over large enough distance scales, then, to a good approximation, the universe can be described as perfectly homogeneous. This assumption implies that our location in space is not in any way special: all places in the universe are more or less the same. Einstein also assumed that the universe is on average *isotropic*, which means that from any point it looks more or less the same in all directions.

Finally, Einstein assumed that the average properties of the universe do not change with time. In other words, the universe is static. Although Einstein had little observational evidence to support this assumption, the picture of an eternal, unchanging universe seemed very compelling.

Having specified the kind of universe he was looking for, Einstein could now try to find a solution of his equations that would describe a universe with the desired properties. It did not take him long, however, to discover that his theory admitted no such solutions. The reason was very simple: masses distributed throughout the universe refused to stay at rest and "wanted" instead to collapse onto one another, because of their gravitational attraction. Einstein was deeply puzzled and perplexed by this situation. After a year of struggle, he decided that the equations of general relativity had to be modified to allow for the existence of a static world.

Einstein realized that it was possible to add an extra term to his equations without violating the physical principles of the theory. The effect of the new term was to endow empty space, or vacuum, with nonzero energy and tension. Each cubic centimeter of empty space has a fixed amount of energy (and therefore mass). Einstein called this constant energy density of the vacuum the *cosmological constant*.* The mathematics of Einstein's equations dictates that the tension of the vacuum is exactly equal to its energy density and is therefore determined by the same constant. The vacuum tension is like the tension in a stretched rubber band that would cause the band to shrink if you let it go. Tension is opposite to pressure, which causes things to expand—as when a balloon expands under the pressure of compressed air. Thus, tension acts as negative pressure.

If the vacuum has energy and tension, how come they seem to have no effect on us? Why don't we see empty space shrink because of its tension? The reason is that it is not so easy to notice *constant* energy and tension. If you increase the pressure inside a balloon, it will expand. But if you also increase the air pressure outside the balloon by the same amount, then there will be no effect. Similarly, if negative-pressure vacuum fills the entire universe, its overall effect is nil. The energy of the vacuum is elusive because it is impossible to extract this energy. You cannot burn the vacuum; you cannot use it to run a car or a hair dryer. Its energy is set by the cosmological constant and cannot be reduced. Thus, the energy and tension of the vacuum are undetectable—except for their gravitational effect.

The gravity of the vacuum turned out to hold a big surprise. According to general relativity, pressure and tension contribute to the gravitational force of massive bodies. If you compress an object, its gravity is enhanced, and if you stretch it, gravity is reduced. This effect is normally very small,

---

*In fact, Einstein did not offer any physical explanation for the new constant. The modern interpretation in terms of the vacuum energy and pressure was later suggested by the Belgian physicist Georges Lemaître.

but if you could keep stretching the object without breaking it, you could in principle reduce gravity to the point of completely neutralizing it, or even making it repulsive. This is precisely what happens in the case of the vacuum. The repulsive gravity of vacuum tension is more than sufficient to overcome the attractive pull of its mass, so the net result is gravitational repulsion.

This property was exactly what Einstein needed to solve his problem. He could now adjust the value of the cosmological constant so that the attractive gravitational force of matter is balanced by the repulsive gravity of the vacuum. The result is a static universe. He found from his equations that the balance is achieved when the cosmological constant is half the energy density of matter.

A striking consequence of the modified equations was that the space of a static universe must be curved, so that it closes in upon itself like the surface of a sphere. A spaceship moving straight ahead in such a closed universe would eventually come back to its starting point. This closed space is called a three-dimensional sphere. Its volume is finite, although it has no boundary.

Einstein described his closed-universe model in a paper published in 1917. He admitted that he had no observational evidence for a nonzero cosmological constant. His only reason for introducing it was to save the static picture of the world. More than a decade later, when the expansion of the universe was discovered, Einstein regretted he had ever proposed the idea and called it the greatest blunder of his life.[3] After this unsuccessful debut, repulsive gravity disappeared from mainstream physics research for nearly half a century—but only to return later with a vengeance.

3

Creation and Its Discontents

As a scientist I simply do not believe that the universe began with a bang.　　　　　　　　　　—SIR ARTHUR EDDINGTON

FRIEDMANN'S UNIVERSES

The cold and hungry Petrograd of the early 1920s was not on anyone's list of places where the next breakthrough in cosmology was likely to occur. Classes at Petrograd University had just resumed, after six years of war and Russian revolution. A young, bespectacled professor was lecturing in a freezing classroom to an audience of students in overcoats and fur hats. His name was Alexander Friedmann. The lectures were meticulously prepared and emphasized mathematical rigor. The courses he taught ranged from mathematics and meteorology, his main areas of expertise, to his most recent passion, the general theory of relativity.

Friedmann was fascinated by Einstein's theory and threw himself into studying it with his usual intensity. "I am an ignoramus," he used to say. "I don't know anything. I have to sleep even less and not allow myself any distractions, because all this so-called 'life' is a complete waste of time."[1] It was as if he knew that he had only a few years left to live—and so much to accomplish.

Having mastered the mathematics of general relativity, Friedmann focused on what he thought was its central problem: the structure of the entire universe. He learned from Einstein's paper that without a cosmological constant, the theory had no static solutions. He wanted to know, however, what kind of solutions it did have. Here, Friedmann made a radical step that would immortalize his name. Following Einstein, he assumed that the universe was homogeneous, isotropic, and closed, having the geometry of a three-dimensional sphere. But he broke away from the static paradigm and allowed the universe to move. The radius of the sphere and the density of matter could now change with time. With the requirement of a static universe lifted, Friedmann found that Einstein's equations do have a solution. It describes a spherical universe that starts from a point, expands to some maximum size, and then recollapses back to a point. At the initial moment, which we now call the big bang, all matter in the universe is packed into a single point, so the density of matter is infinite. The density decreases as the universe expands and then grows as it recontracts, to become infinite again at the moment of the "big crunch," when the universe shrinks back to a point.

The big bang and the big crunch mark the beginning and the end of the universe. Because of the vanishing size and the infinite density of matter, the mathematical quantities appearing in Einstein's equations become ill-defined, and spacetime cannot be extended beyond these points. Such points are called *spacetime singularities*.

A two-dimensional spherical universe can be pictured as an expanding and recontracting balloon (see Figure 3.1). The squiggles on the surface of the balloon represent galaxies, and as the balloon expands, the distances between the galaxies grow. Hence, an observer in each galaxy sees other galaxies rush away. The expansion is gradually slowed down by gravity; it will eventually come to a halt and be followed by the contraction. In the contracting phase, the distances between the galaxies will decrease and all observers will see galaxies moving toward them.

It does not make much sense to ask what the universe is expanding into. We picture the balloon universe as expanding into the surrounding space, but this does not make any difference for its inhabitants. They are confined to the surface of the balloon and are not aware of the third, radial dimension. Similarly, for observers in a closed universe, the three-dimensional spherical space is all the space there is, with nothing else outside it.

•

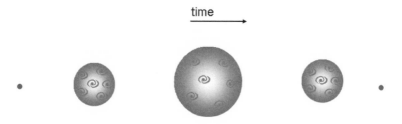

Figure 3.1. An expanding and recontracting spherical universe.

Shortly after publishing these results, Friedmann discovered another class of solutions with a different geometry. Instead of curving back on itself, the space in these solutions curves, in a certain sense, away from itself, resulting in an infinite (open) universe. A two-dimensional analogue of this type of space is the surface of a saddle (Figure 3.2).

Once again, Friedmann found that the distance separating any pair of galaxies in an open universe grows, starting from zero at the initial singularity. The expansion slows down initially, but in this case the force of gravity is not strong enough to turn it around, so at late times galaxies move apart at nearly constant speeds.

On the borderline between the open and closed models is the universe

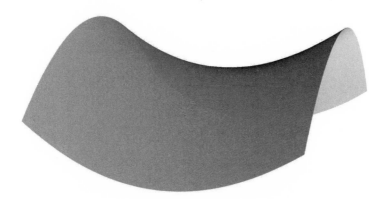

Figure 3.2. A two-dimensional analogue of an open universe.

with a flat, Euclidean geometry.[2] It expands without limit, but barely so, with the expansion rate becoming smaller and smaller with time.

A remarkable feature of Friedmann's solutions is that they establish a connection between the geometry of the universe and its ultimate fate. If the universe is closed, it must recollapse, and if it is open or flat, it will expand forever.* In his papers Friedmann gave no preference to either model.

Unfortunately, Friedmann did not live to see his work become the foundation of modern cosmology. He died of typhoid fever in 1925, at the age of thirty-seven. Although Friedmann's papers were published in a leading German physics journal, they went almost unnoticed.[3] His papers were unearthed in the 1930s, in the wake of Hubble's discovery of the expansion of the universe.†

THE MOMENT OF CREATION

Whatever Friedmann's solutions have to say about the future of the universe, their most unexpected and intriguing aspect is the presence of the initial singularity—the big bang, where the mathematics of general relativity breaks down. At the singularity, matter is compressed to infinite density, and the solutions cannot be extended to earlier times. Thus, taken literally, the big bang should be interpreted as the beginning of the universe. Was that the creation of the world? Could it be that the whole universe began in a singular event a finite time ago?

For most physicists this was too much to take. A singular jump-starting of the universe looked like a divine intervention, for which they thought there should be no place in physical theory. But although the "beginning of the world" was—and to a large degree remains—a source of discomfort for most scientists, it also had some benefits to offer. It helped to resolve some perplexing paradoxes that haunted the picture of a static, eternally unchanging universe.

To begin with, an eternal universe appears to be in conflict with one of

*The simple connection between the geometry of the universe and its fate holds only assuming that the vacuum energy density (or cosmological constant) is equal to zero. More on this in Chapter 18.

†The expanding universe model was reinvented in 1927 by Georges Lemaître. Just like Friedmann's work, Lemaître's paper remained completely unknown until Hubble's discovery.

the most universal laws of nature: the second law of thermodynamics. This law says that physical systems evolve from ordered to more disordered states. If you neatly organize papers into piles on your desk and the wind suddenly blows into the window, the papers are scattered randomly all over the floor. However, you never see the wind picking up papers from the floor and assembling them into neat piles on your desk. Such a spontaneous reduction of disorder is not impossible in principle, but it is so unlikely that it is never seen to occur.

Mathematically, the amount of disorder is characterized by the quantity called *entropy*, and the second law says that the entropy of an isolated system can only increase. This relentless increase of disorder leads eventually to the state of maximum possible entropy, *thermal equilibrium*. In this state all the energy of ordered motion has been turned into heat and a uniform temperature has been established throughout the system.

The cosmic implications of the second law were first pointed out in the mid-1800s by the German physicist Hermann von Helmholtz. He argued that the whole universe can be regarded as an isolated system (since there is nothing external to the universe). If so, then the second law is applicable to the universe as a whole, and thus the universe should be headed toward an inevitable "heat death" in the state of thermal equilibrium. In that state the stars will all be dead and have the same temperature as their surroundings, and all motion will come to a halt, other than the disordered thermal jostling of the molecules.

Another consequence of the second law is that if the universe existed forever, it should have already reached thermal equilibrium. And since we do not find ourselves in the state of maximum entropy, it follows that the universe could not have existed forever.[4]

Helmholtz did not emphasize this second conclusion and was more concerned about the "death" part (which by the way inspired much apocalyptic prose in the late nineteenth and early twentieth centuries). But other physicists, including giants like Ludwig Boltzmann,* were well aware of the problem. Boltzmann saw the way out in the statistical nature of the second law. Even if the universe *is* in the maximally disordered state, spontaneous reduc-

---

*Boltzmann established the connection between entropy and disorder and elucidated the meaning of the second law.

tions of disorder will occasionally happen by chance. Such events, called *thermal fluctuations*, are common on the microscopic scale of a few hundred molecules, but become increasingly unlikely as you move toward larger scales. Boltzmann suggested that what we are observing around us is a huge thermal fluctuation in an otherwise disordered universe. The probability for such a fluctuation to happen is unbelievably small. However, improbable things do eventually happen if you wait long enough, and they will definitely happen if you have infinite time at your disposal. Life and observers can exist only in the ordered parts of the universe, and this explains why *we* are observing this incredibly rare event.[5]

The problem with Boltzmann's solution is that the ordered part of the universe appears to be excessively large. For observers to exist, it would be enough to turn chaos into order on the scale of the solar system. This would have a much higher probability than a fluctuation on the scale of billions of light-years that would be needed to account for the observed universe.

Another problem, having an even longer pedigree, arises if one assumes that the universe is infinite and that stars (or galaxies) are distributed more or less uniformly throughout the infinite space. If this were the case, then no matter where you looked in the sky, your line of sight would eventually hit upon a star. The sky would then constantly glare with a nearly uniform brilliance—which leaves us with a simple question: Why is it dark at night? The problem was first recognized in 1610 by Johannes Kepler, whose conclusion was that the universe could not be infinite.

Both the entropy problem and the night sky paradox are naturally resolved if the age of the universe is finite. If the universe came into being only a finite time ago and was initially in a highly ordered (low entropy) state, then we are now observing the descent from that state into chaos and should not be surprised that the state of maximum disorder has not yet been reached. The night sky paradox is resolved because, in a universe of a finite age, light from very distant stars has not had enough time to reach us. We can only observe the stars within the *horizon* radius, equal to the distance traveled by light during the lifetime of the universe. The number of stars within that radius is clearly finite, even if the entire universe is infinite.

Given these arguments, how could anyone ever believe that the universe as we know it has existed forever? The reason is, of course, that the idea of a cosmic beginning that occurred a finite time ago creates perplexing problems of its own. If the universe began a finite time ago, then what determined the initial conditions at the big bang? Why did the universe start in a homogeneous and isotropic state? It could in principle start in any state at all. Should we attribute the choice of the initial state to the Creator? Not surprisingly, physicists were slow to embrace the big bang cosmology and made numerous attempts to avoid dealing with the problem of "the beginning."

ACCEPTING THE INEVITABLE

Some people initially suggested that the big bang singularity was an artifact of the assumptions of exact homogeneity and isotropy that Friedmann adopted to solve Einstein's equations. In a collapsing universe, if all galaxies were moving radially toward us, it would be no wonder that they would all crush together in a big crunch. But if the motion of galaxies were even slightly nonradial, one might think that they would bypass one another and start flying apart afterward. The singularity would then be avoided, and contraction would be followed by an expansion. Thus, one might hope to construct an oscillating model of the universe, without a beginning, with alternating periods of expansion and contraction.

It turns out, however, that the attractive nature of gravity makes this scenario impossible. The British physicist Roger Penrose and Stephen Hawking, who was a graduate student at the time, proved a series of theorems showing, under very general assumptions, that the cosmological singularity cannot be avoided. The main assumptions used in the proofs are that Einstein's general theory of relativity is valid, and that matter has positive energy density and pressure everywhere in the universe. (More precisely, the pressure should not get so negative as to make gravity repulsive.) Thus, as long as we stay within the framework of general relativity and do not assume exotic repulsive-gravity matter, the singularity will be with us and the question of the initial conditions will remain unresolved.

The most notorious attempt to avoid the problem of the beginning was no doubt the steady-state theory, suggested in 1948 by the British astrophysicist Fred Hoyle and two Austrian refugees, Hermann Bondi and Thomas Gold, all at Cambridge University. They boldly asserted that the universe has always remained unchanged in its broad features, so that it looks more or less the same at all places and at all times. This view seems to be in glaring contradiction with the expansion of the universe: If the distances between the galaxies grow, how can the universe remain unchanged? To compensate for the expansion, Hoyle and his friends postulated that matter is being continuously created out of the vacuum. This matter fills the voids opened by the receding galaxies, so that new galaxies can be formed in their place.

The Cambridge physicists admitted that they had no evidence for the spontaneous creation of matter, but the required creation rate was so low— a few atoms per cubic mile per century—that there was no evidence against it either. They further defended their theory by pointing out that continuous creation of matter, in their view, was no more objectionable than creation of all matter at once in the big bang. In fact, the term "big bang" was coined by Hoyle as he ridiculed the competing theory in a popular BBC radio talk show.

It did not take long, however, for the steady-state theory to run into serious problems. The most distant galaxies are seen as they were billions of years ago, because that is how long it takes for their light to reach us. If the steady-state theory is correct, and the universe at that time was the same as it is now, then these distant galaxies should look more or less the same as the galaxies we now see in our own neighborhood. With more data, however, it became increasingly clear that far-away galaxies are actually quite different and show distinct signs of their youth. They are smaller, have irregular shapes, and are populated with very bright, short-lived stars. Many of them are powerful sources of radio waves, a trait much less common among the older, nearby galaxies.[6] There seemed to be no way in which the observations could be explained in terms of the steady-state theory.

As Sherlock Holmes used to say, "When you have eliminated the impossible, whatever remains, however improbable, must be the truth."[7] The prospects of the steady-state theory were getting dimmer, and with no other viable alternative in sight, attitudes began to shift. Physicists were gradually coming to terms with the picture of an evolving universe that started with a bang.

· 4 ·

The Modern Story of Genesis

*The elements were cooked in less time than it takes to cook a
dish of duck and roast potatoes.* —GEORGE GAMOW

TUNNELING THROUGH THE IRON CURTAIN

The idea of the primeval fireball was born in the mind of George Gamow, a flamboyant Russian-born physicist whom we shall encounter more than once as our story develops. A fellow physicist, Leon Rosenfeld, described him as "a Slav giant, fair haired and speaking a very picturesque German; in fact he was picturesque in everything, even in his physics."[1] Gamow took Friedmann's course in general relativity in 1923–24, while he was a graduate student in Petrograd; thus he heard about the expanding universe solutions, so to say, from the horse's mouth. He wanted to do research in cosmology under Friedmann, but this plan was ruined by Friedmann's sudden death. Gamow ended up writing his thesis on the dynamics of a pendulum, a subject he characterized as "extremely dull."[2]

In 1928, at the instigation of his old professor, Orest Khvolson, Gamow was given a stipend to spend the summer at the University of Göttingen in Germany. That was the time when quantum mechanics was being developed, and Göttingen was one of the leading centers in this area of research. Physicists were trying to capture the essence of the new theory and to contribute to its rapid advance. Discussions that started in seminar rooms during the day continued in the streets and cafés in the evenings, and it was hard not to be infected by this atmosphere of excitement and discovery. Gamow decided to investigate what quantum mechanics could say about the structure of atomic nuclei, and very quickly he made his mark. He used what is called the *tunneling* effect—the penetration of a barrier by a quantum particle—to explain the radioactive decay of nuclei. His theory was in beautiful agreement with the experimental data.

When the summer came to an end and it was time to return to Petrograd (now called Leningrad), Gamow decided to make a stop in Denmark and visit the legendary Niels Bohr, one of the founders of the quantum theory. He told Bohr about his work on radioactivity (which was not yet published), and Bohr was sufficiently impressed to offer Gamow a fellowship at his institute in Copenhagen. Of course, Gamow accepted with enthusiasm. He continued work in nuclear physics and soon became a recognized authority in this field.

In 1930 Gamow was invited to give a major talk at the International Congress on Nuclear Physics in Rome. He was already preparing to cross Europe on his little motorcycle when he learned from the Soviet embassy that his passport could not be extended and that he had to return to the Soviet Union before traveling anywhere else.

Back in Leningrad, Gamow immediately sensed that things had taken a drastic turn for the worse. The Stalinist regime was tightening its grip on the country. Science and art had to conform to the official Marxist ideology, and anyone accused of "bourgeois" idealistic views was severely persecuted. Quantum mechanics and Einstein's theory of relativity were declared nonscientific and contrary to Marxism-Leninism. When Gamow mentioned quantum physics in a public lecture, a government representative interrupted the lecture and dismissed the audience. Gamow was warned that such mistakes were not to be repeated. Even before this incident, he was told he could forget about foreign travel and should not bother applying for a passport. The iron curtain was tightly closed. In Gamow's mind, the writing was on the wall: he had to escape from the Soviet Union.

With his wife Lyuba, whom he had married soon after his return to Leningrad, Gamow was preparing for the escape. The plan was to cross the Black Sea from the Crimean Peninsula to Turkey. Childish as it may seem, they wanted to do this in a kayak. They had a food supply for a week and a simple navigation plan: paddling straight to the south. But the Black Sea is not called black for nothing. Perfectly calm when the two adventurers left in the morning, the sea became increasingly rough toward the evening. During the night, it took all their efforts to keep the boat from turning over. Accepting defeat, they were now fighting to get back to the shore and felt fortunate when they finally made it the following day.

It was totally unexpected when in the summer of 1933 Gamow was informed that he had been appointed to represent the Soviet Union at the prestigious Solvay Congress on nuclear physics in Brussels. He was overjoyed, but had no idea what to make of it. The explanation came on arrival at the congress. When Gamow did not show up in Rome, Niels Bohr got concerned and wanted to see his old friend. He asked the French physicist Paul Langevin, a member of the French Communist Party, to use his connections to arrange Gamow's appointment to the Solvay Congress. But, Gamow was horrified to find out, Bohr gave Langevin his personal assurance that Gamow was going to return to the Soviet Union! That evening at the dinner table Gamow sat next to Marie Curie, the famous discoverer of radium and plutonium, and told her about his impossible situation. Madame Curie knew Langevin very well (rumors said too well); she said she would talk to him. After a sleepless night and a day of anxious anticipation, Gamow finally heard from Curie that the issue was settled and he did not have to go back. The following year he accepted a professorship at George Washington University in the United States.

THE PRIMEVAL FIREBALL

Gamow realized that the early universe was not only superdense, it was also superhot. The reason is that gases get hotter when they are compressed and cool down when they expand. (People who ride bicycles tell me that they know this property firsthand: a bicycle tire gets warm when you pump it with air. The compressed air heats up and the surface of the tire gets warmer as a result.)

To see why expansion causes a gas to cool down, consider a gas contained in a large box. You can picture the gas molecules as little balls bouncing off the walls of the box. Imagine now that the walls are moving apart, so that the box is expanding. What effect will the recession of the walls have on the molecules? If you hit a tennis ball against a wall during a tennis practice, the ball comes back at you at the same speed. But imagine for a moment that the wall is moving away from you. The ball's speed relative to the wall would then be smaller, and it would bounce back slower than you sent it off. Similarly, the molecules in an expanding box will slow down on each reflection from the walls. The temperature is proportional to the average energy of the molecules and will therefore decrease in the course of expansion. Of course, there are no moving walls in the expanding universe, but particles are reflected off one another, and the effect on the temperature is the same. The universe was getting progressively colder as it expanded. Thus, if we go back in time, the universe gets hotter and hotter, and it becomes infinitely hot if we extrapolate all the way back to the singularity.

At temperatures above a few hundred degrees kelvin,* the bonds holding atoms together inside molecules are not strong enough to withstand the heat, and the molecules decompose into separate atoms. Further increase of temperature leads to a progressive breakup of atoms. First, at about 3000 degrees kelvin, electrons are stripped off the atomic nuclei,[3] then at a billion degrees or so the nuclei fragment into protons and neutrons (collectively called *nucleons*), and finally at about a trillion degrees the nucleons break apart into their elementary constituents, called *quarks*.

Apart from matter particles that make up atoms, the fireball also contained vast quantities of radiation quanta, called *photons*. Photons are bundles of electric and magnetic energy; they are what ordinary visible light is made of. Moving charged particles emit and absorb photons, so equilibrium is quickly established where photons are absorbed at the same rate as they are emitted. The higher the temperature is, the higher are the average energy and the density of photons in equilibrium. The recipe for the hot cosmic soup thus appears to be very simple: break everything down to the smallest pieces, and then mix together and add a suitable quantity of photons. But there is more to it than that.

The further back in time we go, the more energetic the particles become. They are also more densely packed, and constantly bump into one another.

---

*In the kelvin scale, often used by physicists, the temperature is measured in centigrade units starting from absolute zero (−273.15 degrees Celsius). For the very high temperatures we are discussing here, there is little difference between the Celsius and kelvin scales.

To understand the makeup of the fireball, we need to know what happens in such high-energy collisions. Smashing elementary particles is the favorite occupation of particle physicists. They build monstrous machines, called particle accelerators, where they boost particles to huge energies, let them collide, and see what happens. This is much more exciting than watching billiard balls collide, because particles often change their identity in collisions—it would be as if red and blue balls turned into yellow and green ones as they hit one another. The number of particles can also be altered: two initial particles can produce fireworks with dozens of new particles flying away from the collision point. This type of event was commonplace in the early moments after the big bang.

In such a collision, you cannot predict exactly what is going to happen. There is a large number of possible outcomes, and physicists use quantum theory to calculate their probabilities. But this is as far as you can go: there is no certainty in the quantum world. The range of possibilities is constrained by a few *conservation laws*, which are strictly enforced. Examples are energy and charge conservation: the total energy and the total electric charge should be the same before and after collision. Any process that is not forbidden by the conservation laws is thereby allowed and will occur with some nonzero probability. In the early universe, particles are incessantly hitting one another, and the fireball gets populated with all types of particles that can be created in these encounters.

For each type of particle, there exists an antiparticle of precisely the same mass and opposite electric charge. Particles and antiparticles are often created in pairs. For example, two photons with energies greater than that associated with the electron mass (through $E = mc^2$) can collide and turn into an electron and its antiparticle, called a *positron*. The opposite process is *pair annihilation*: an electron and a positron smash into one another and turn into two photons.

At temperatures above 10 billion degrees, particle energies become large enough to produce electron-positron pairs. As a result, the fireball gets populated with a gas of electrons and positrons having about the same density as the gas of photons. At still higher temperatures, pairs of increasingly heavier particles make their appearance. Physicists have catalogued an extensive zoo of particles with a wide range of masses. At the top of this range are W and Z particles, which are about 300,000 times more massive than electrons, and the *top quark*, about twice as heavy as W or Z. These are the heaviest particles that can currently be produced in particle accelerators. They existed in the fireball at temperatures above 3000 trillion degrees. As we approach these temperatures, our knowledge of particle physics becomes more and more sketchy and our understanding of the primeval fireball more and more uncertain.

Friedmann's equations can be used to determine what temperature and density the fireball had at any given time. For example, at 1 second after the big bang, the temperature was 10 billion degrees and the density about 1 ton per cubic centimeter. (To avoid repeating "after the big bang," I will use the abbreviation "A.B.")

The most eventful part of the fireball history, marked by a rapid succession of exotic particle populations, occurred during the first second of its existence. The W, Z, and heavier particles were abundant only in the first 0.00000000001 second A.B. Muons—particles similar to electrons but 200 times heavier—and their antiparticles annihilated at 0.0001 second. At about the same time, triplets of quarks merged together to form nucleons. The last to annihilate were electron-positron pairs. They disappeared at 1 second A.B. There must have been a slight excess of quarks over antiquarks and of electrons over positrons to leave us with some electrons and nucleons at present.[4] After the first second, the remaining components of the cosmic soup were nucleons, electrons, and photons.*

GAMOW'S ALCHEMY

Particles like quarks or W and Z were not known in Gamow's day, and he was not even concerned about electron-positron pairs. His main interest was in the cosmic history after 1 second A.B. Early in his career Gamow became fascinated with the problem of the origin of atoms. There are ninety-two different types of atoms, or chemical elements, found in nature. Some of them, like hydrogen, helium, and carbon, are very abundant, while others, like gold and uranium, are extremely rare. Gamow wanted to know why this is so: What determined the element abundances?

In the Middle Ages alchemists tried to turn more abundant elements

---

*Also present in the fireball were very light, weakly interacting particles called neutrinos.

into gold, but as we now know, there was a good reason why they did not succeed. In order to change one chemical element into another, one has to learn how to change the composition of atomic nuclei. But the particle energies needed for such nuclear transformations are millions of times greater than the energies typically involved in chemical reactions, far beyond what alchemists could achieve. Such energies are reached in a hydrogen bomb, but are not attained in any process naturally occurring on Earth. Thus, the element abundances we observe now are the same as they were 4.6 billion years ago, when the solar system was formed.*

A natural place to look for the origin of elements is in the interiors of stars. Stars are giant, hot, gaseous spheres held together by gravity. Our Sun consists mainly of hydrogen—the simplest element whose nucleus is made of a single proton. The temperature in the central regions of the Sun is higher than 10 million degrees, high enough for nuclear reactions to occur. A chain of reactions transforms hydrogen into helium, releasing the energy that fuels the Sun. The theory of nuclear reactions in the Sun was developed in the late 1930s by Hans Bethe, a German-born physicist who was later awarded a Nobel Prize for this work. This theory, however, did very little to explain the elemental abundances. Helium production in stars can account for only a small fraction of the vast amounts of helium observed in the universe. Another puzzle is the presence of deuterium (heavy hydrogen), which has a very fragile nucleus. Deuterium is quickly destroyed in hot stellar interiors, and it is hard to see how it could ever be produced.

Gamow's assessment of the situation was that stars were simply not hot enough to cook the elements, and he thought he had a better idea of what a suitable furnace could be: the entire universe shortly after the big bang. To investigate nuclear processes in the hot, early universe, Gamow enlisted the help of two young physicists, Ralph Alpher and Robert Herman. They considered a hot mixture of nucleons, electrons, and radiation uniformly filling the universe. When the universe cools down below 1 billion degrees kelvin, it becomes possible for a neutron and a proton to stick together and form a nucleus of deuterium (see Figure 4.1). Further attachment of protons and

---

*Radioactive elements, such as uranium, which spontaneously decay into lighter elements, are an important exception. A uranium atom decays into lead with an average lifetime of 4.5 billion years, so the amount of uranium is gradually decreasing. In fact, our best estimate for the age of the Earth comes from measurements of the relative abundances of uranium and lead.

neutrons quickly turns deuterium into helium (which has two protons and two neutrons in its nucleus). At this point, however, the buildup of nuclei essentially stops. The reason is that owing to some peculiarity of nuclear forces, there are no stable nuclei consisting of five nucleons, and simultaneous attachment of more than one nucleon is highly unlikely. This is what's known as the five-nucleon gap. Calculations show that about 23 percent of all nucleons end up in helium, and almost all the rest in hydrogen. Small amounts of deuterium and lithium are also produced.[5]

Modern analyses, using the latest data on nuclear reactions and extensive computer power, give precise element abundances as they come out of the cosmic furnace. These calculations are in a very impressive agreement with astronomical observations. By studying the spectrum of light emitted by distant objects, astronomers can determine their chemical composition. A firm prediction of the hot big bang theory is that no galaxy in the universe should contain less than 23 percent of helium: helium produced in stars can only increase this primordial abundance. And indeed, no such galaxy has yet been found. The predicted abundance of deuterium is somewhat less than one part in 10,000, and the abundance of lithium is less than one part in a billion. It is quite remarkable that these vastly different amounts are con-

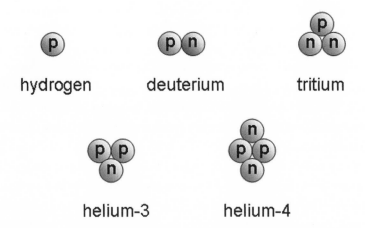

Figure 4.1. Simplest atomic nuclei, with protons and neutrons represented by p and n, respectively.

firmed by observations. You might say that 23 percent of helium was a lucky guess, but the probability of a chance coincidence for the whole set of numbers is extremely low.

But what about the heavy elements? Despite all their efforts, Gamow and his crew could not find a way to bridge the five-nucleon gap. In the meantime, across the Atlantic, the chief proponent of the steady-state model, Fred Hoyle, was developing an alternative theory for the origin of elements. Hoyle was aware that stars, like our Sun, that burn hydrogen into helium are not hot enough to do the job. But what happens when a star runs out of its hydrogen? Then it can no longer support itself against gravity, so the stellar core begins to contract, with its density and temperature rising. When the central temperature reaches 100 million degrees, a new channel of nuclear reactions opens up: three helium nuclei stick together to form a nucleus of carbon. When all the helium in the central region is consumed, the star contracts further, until the temperature gets high enough to ignite carbon-burning nuclear reactions. As the process continues, a layered structure is formed with heavier elements closer to the center (since they require higher temperatures to be cooked). The process does not get very far in stars like the Sun, but in more massive stars it goes all the way to iron, which is the most stable of all nuclei. Beyond that, there is no more nuclear fuel to burn. Unsupported by nuclear reactions, the innermost core of the star collapses, reaching enormous densities and temperatures as high as 10 billion degrees. This triggers a gigantic explosion, known as a *supernova*, when all outer layers of elements are blown off into the interstellar space. Elements heavier than iron are formed during the core collapse and explosion. The enriched interstellar gas serves as a raw material for new stars and planetary systems. The resulting abundances of heavy elements, calculated by Hoyle and his collaborators, are in good agreement with observations.

Hoyle and Gamow were developing their ideas in the 1940s and '50s, and at the time their theories were regarded as two competing models for the origin of elements. But in the end they both turned out to be right: light elements were formed predominantly in the early universe, and heavy elements in stars. Almost all known matter in the universe is in the form of hydrogen and helium, with heavy elements contributing less than 2 percent.

However, the heavy elements are crucial for our existence: Earth, air, and our bodies are made mostly of the heavy elements. As Cambridge astrophysicist Martin Rees wrote, "We are stardust—the ashes of long-dead stars."[6]

COSMIC MICROWAVES

The process of helium formation began at about 3 minutes A.B. and was complete in less than a minute. The universe was still expanding at a furious rate, and both the density and the temperature were dropping very rapidly. But after the first few minutes packed with action, the pace of the cosmic drama was getting slower. Very little was happening to the matter particles, and the most notable change was in the radiation component of the fireball.

At the microscopic, quantum level, the radiation consists of photons, but macroscopically it can be pictured as consisting of electromagnetic waves—oscillating patterns of electric and magnetic energy. The higher the frequency of oscillation, the more energetic the constituent photons. Waves of different frequency produce different physical effects, and we know them under different names. Visible light corresponds only to a narrow range of frequencies in the full electromagnetic spectrum. Higher frequency waves are called X rays, and still higher-frequency waves are called gamma rays. Going down in frequency, we encounter microwaves and still lower, radio waves. All these waves propagate at the speed of light.

As the fireball temperature declined, the intensity of the radiation tapered, and its frequency gradually shifted from gamma rays to X rays and then to visible light. An important event occurred at 300,000 years A.B., when the temperature got low enough for electrons and nuclei to combine into atoms. Prior to that, electromagnetic waves were frequently scattered by charged electrons and nuclei. However, the interaction of radiation with electrically neutral atoms is very weak, so that once atoms were formed, the waves propagated freely through the universe, with practically no scattering at all. In other words, the universe suddenly became transparent to light.

What happens to the cosmic radiation after that? Not much, except the frequencies of the electromagnetic waves, and the corresponding temperature, keep declining with the expansion of the universe. At the time of neutral atom formation, the temperature of the radiation was 4000 degrees, somewhat below that at the surface of the Sun. If we had been there, and could have tolerated such unhealthy conditions, we would have seen the universe ablaze with brilliant orange light. By the cosmic age of 600,000 years the light would change to red. At 1 million years, it would shift beyond the visible range, to the infrared part of the spectrum. So, as far as we'd have been concerned, the universe would have descended into complete darkness. Wave frequencies still continue to decline slowly: by the present time, corresponding to the cosmic age of about 14 billion years, they are down to the microwave range.

This history of the cosmic fireball was studied by Alpher and Herman, Gamow's young collaborators. They followed it all the way to the present and reached a remarkable conclusion—that we should now be immersed in a sea of microwaves having the temperature of about 5 degrees kelvin.

Alpher and Herman's work was published in 1948. You might think that it should have inspired a fair number of observers to search for cosmic microwaves. Indeed, the primeval radiation is a true smoking gun of the big bang, and its discovery should have a colossal significance. You might think also that, once the radiation is detected, a Nobel Prize would be awarded for its prediction. Alas, this is not how the events unfolded.

THE SMOKING GUN

Odd as it may seem, the prediction of cosmic radiation was completely ignored for nearly two decades, until the radiation was accidentally discovered in 1965. Two radio astronomers, Arno Penzias and Robert Wilson, working at Bell Telephone Laboratories in New Jersey, detected a persistent noise in their sensitive radio antenna. The noise level could be characterized by a temperature of approximately 3 degrees kelvin and did not depend on the time of day or on the direction in which they pointed the antenna. Determined to get to the root of the problem, Penzias and Wilson painstakingly eliminated all possibilities they could think of. This included eviction of a pair of pigeons who were roosting in the antenna and removing what Penzias called the "white dielectric material" that was left after them. Nothing worked, however, and the origin of the noise remained enigmatic.

In the meantime, about 30 miles away, a group of physicists at Princeton University were busy building a radio detector of their own. The head of the group was Robert Dicke, an extraordinary physicist who was equally at home in theory and experiment. Dicke realized that a hot early stage in the history of the universe should have left an afterglow, and he designed an antenna to search for it. When the Princeton group were ready to start their measurements, they learned about Penzias and Wilson's predicament. They knew immediately that the bothersome noise that Penzias and Wilson were working so hard to eliminate was precisely the signal of cosmic microwaves that they were hoping to detect!

It is a fascinating question why the cosmic radiation had to be discovered by accident. Why had nobody listened to Alpher and Herman? Even if their papers were somehow overlooked, why did it take more than fifteen years for someone else to come up with the same prediction? After all, cosmic radiation was a direct consequence of Gamow's hot big bang model.

One reason, it seems, was that physicists simply did not believe that the early universe was for real. "This is often the way it is in physics," wrote the Nobel Prize–winning physicist Steven Weinberg. "Our mistake is not that we take our theories too seriously, but that we do not take them seriously enough."[7] It did not help also that George Gamow was perhaps too colorful a character to be taken seriously by the physics community. A practical joker, composing "unprintable" limericks and often having one too many at the bar, he was surely not your typical physicist. Finally, by the mid-1950s neither Gamow nor Alpher and Herman were actively working on the big bang theory: Gamow was increasingly attracted to biology, where he suggested important insights into the genetic code, while Alpher and Herman left academia and moved on to careers in private industry. One cannot help thinking that lack of appreciation of their work must have played a role in those decisions. By the mid-1960s, when Penzias and Wilson were taking data from their antenna, the work of the Gamow group was all but forgotten.

Penzias and Wilson measured the intensity of the radiation at a single frequency (to which their antenna was tuned), while the theory predicted that the radiation should be spread over a range of frequencies, with the intensity following a simple formula derived by Max Planck at the turn of the twentieth century. This prediction was spectacularly confirmed in 1990 by the Cosmic Background Explorer (COBE) satellite experiment, which found agreement with the Planck formula at the level of one part in 10,000.

The discovery of the cosmic radiation was no doubt an epoch-making event in cosmology. This tangible relic of the primeval fireball gives us faith that we have not dreamed it all up, that there was indeed a hot early universe some 14 billion years ago. Penzias and Wilson received the 1978 Nobel Prize "for their discovery of cosmic microwave background radiation." No prize for its theoretical prediction has ever been awarded.

IMPERFECTIONS OF CREATION

If the universe had started out perfectly homogeneous, then it would remain homogeneous to this day. The thin, uniform gas filling the universe would gradually be getting ever thinner, and the universe would remain permanently dark, with cosmic radiation slowly shifting to radio waves of lower and lower frequency. But one look at the night sky should be enough to convince you that our universe is not nearly so dull. The universe is lit up with shining stars that are scattered throughout space, forming a hierarchy of structures. The basic unit of this hierarchy is the galaxy, with a typical galaxy containing about 100 billion stars. Galaxies are grouped in clusters, which in turn form superclusters that extend up to a few hundred million light-years*—only 100 times smaller than the size of the currently observable universe.

Cosmologists attribute the origin of all these magnificent structures to tiny inhomogeneities that existed in the primeval fireball. Small inhomogeneities can grow into galaxies as a result of *gravitational instability*. Suppose some region of the universe is slightly denser than its surroundings. It will have stronger gravity and will attract more matter from the surrounding regions. As a result, the density contrast will keep growing, and a nearly homogeneous initial distribution of matter will evolve into a highly inhomogeneous one. Cosmologists believe that this is how galaxies, clusters, and superclusters were formed. According to the theory, the first galaxies were formed about 1 billion years after the big bang. Stars lit up the universe, and thus the cosmic dark age ended. The process of galaxy formation was complete in the not-so-distant past—at the cosmic age of about 10 billion years ("only" 4 billion years ago).

*See footnote on p. 14 for a definition of "light-year."

You might think that this story is destined to remain just that—a story—since nobody was there to confirm it. But as I already emphasized, we see distant objects as they were a long time ago, when the light we now detect was emitted. Thus, by studying more distant galaxies, we go further back in time. The travel time of light from the most distant galaxies that we can observe is about 13 billion years, so we see them when the universe was a billion years old. Compared to the grand spirals we find nearby, these galaxies are small and irregular—a sign of their youth.

Still earlier epochs in the history of the universe can be observed through cosmic microwaves. These waves traveled without scattering for nearly 14 billion years, since the time when the universe became transparent to radiation. The regions where the waves were last scattered are now 40 billion light-years away* (not 14 billion light-years as one might think, since the universe was expanding in the meantime). Thus, the microwaves come to us from the surface of a gigantic sphere, 40 billion light-years in radius; it is called the *surface of last scattering*. Radiation emitted from regions of slightly higher density has to overcome stronger gravity and arrives to us with a slightly diminished intensity. As a result, denser regions look dimmer on the microwave sky. By mapping the radiation intensity from different directions in the sky, we can obtain an image of the universe at the epoch of last scattering, when it was only 300,000 years old.

The first successful map of the microwave sky was made by the COBE team in 1992. A more detailed map, produced 10 years later by the WMAP satellite,† is shown in Figure 4.2. Darker shades of grey correspond to higher radiation intensity, but the difference in intensity between the lightest and darkest spots is only a few parts in 100,000. This means that at the time of last scattering the universe was almost perfectly homogeneous. All the glorious structures that we now see in the sky were then encoded in tiny amorphous ripples on the smooth cosmic background.

---

*We say that electromagnetic waves are scattered when they are absorbed and re-emitted by charged particles. The last scattering surface could therefore be equally well characterized as the surface where the cosmic radiation was emitted.

†The Wilkinson Microwave Anisotropy Probe, so named after David Wilkinson of Princeton University. Wilkinson originated the idea of the probe and was a major inspiration in its design. Sadly, he died shortly before the satellite was launched.

Figure 4.2. Microwave sky as mapped by the WMAP satellite. (Courtesy of Max Tegmark)

THE MODERN STORY OF GENESIS

The picture in Figure 4.3 illustrates the story of genesis as we have discussed it so far. This story is supported by an abundance of observational data, and there is little doubt that it is basically correct. The details are still being worked out, and some outstanding questions remain open. One of the big unknowns is the nature of the dark matter that manifests itself by its gravitational pull in galaxies and clusters. There are strong reasons to believe that most of this dark matter is not made up of nucleons and electrons, but rather consists of some yet undiscovered particles. The details of the galaxy formation process depend on the masses and interactions of these particles, but the general picture outlined in Figure 4.3 does not.

It is truly remarkable that we can observe the universe as it was 14 billion years ago and accurately describe the events that took place a fraction of a second after the big bang. This brings us tantalizingly close to the moment of creation. But what actually happened at that moment remains as enigmatic as ever. In fact, on closer examination the big bang turns out to be even more peculiar than it seemed before.

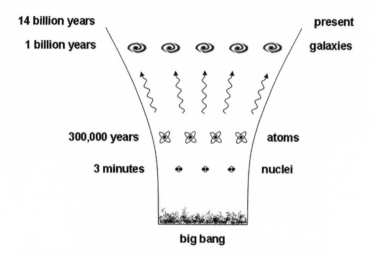

Figure 4.3. Abridged history of the universe.

5

The Inflationary Universe

An invasion of armies can be resisted, but not an idea whose time has come. —Victor Hugo

COSMIC PUZZLES

Suppose one day you receive a radio message from a distant galaxy saying "Elvis lives." You point your antenna to a different galaxy, and to your surprise you get an identical message! Mystified, you turn the antenna from one galaxy to another, but the same message keeps coming to you from all over the sky. One conclusion that you draw is that the universe is full of Elvis fans; the other is that they are in communication with one another. How else would they come up with identical messages?

Silly as it is, this example closely resembles the situation we observe in our universe. The intensity of the microwave radiation coming to us from all directions in the sky is the same, with a very high degree of accuracy, which indicates that the density and temperature of the universe were highly uniform at the time when the radiation was emitted. This observation suggests that there was some interaction between the radiation-emitting regions that led to equilibration of densities and temperatures. The problem is, however, that the time elapsed since the big bang is too short for such an interaction to have occurred.

The crux of the problem is that physical interactions cannot propagate faster than the speed of light. The distance traveled by light since the big bang, about 40 billion light-years, is the *horizon distance*. It puts a limit on how far we can see in the universe and gives the maximum range over which communications could be established. The cosmic radiation that we now observe was emitted shortly after the big bang and comes to us from distances approximately equal to the horizon. Now, consider the radiation coming from two opposite directions in the sky (Figure 5.1). The regions where this radiation was emitted are now separated by twice the horizon distance, and thus could not possibly interact. In particular, they could not exchange heat to equalize their temperature.

At earlier times the two regions were closer to one another, and you might think this could have helped them to equilibrate. But actually at early times the difficulty is even more severe. The reason is that as we go back in time, the horizon distance shrinks even faster than the separation between the regions. At the time of last scattering, when the radiation was emitted, the observable part of the universe was fragmented into thousands of small regions that could not "talk" to one another. We are thus driven to the conclusion that no physical process could make the fireball uniform if it was not uniform to begin with.

This mysterious feature of the big bang is often referred to as the *horizon problem*. The only explanation we can give to the remarkable uniformity of density and temperature in the early universe is that this is how the infant universe emerged from the big bang. Logically, there is nothing wrong with

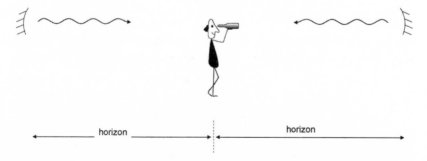

Figure 5.1. Cosmic radiation coming from opposite directions in the sky originated in regions that are now separated by twice the horizon distance.

this "explanation." The physical conditions at the singularity are undetermined, so one can postulate any physical state immediately after the big bang. But one cannot help feeling that this does not explain anything at all.

Another puzzling feature of the big bang is the precarious balance between the power of the blast that sent all particles rushing away from one another and the force of gravity that slows the expansion down. If the density of matter in the universe were a bit higher, its gravitational pull would be strong enough to halt the expansion and the universe would eventually recollapse. If it were a bit lower, the universe would continue expanding forever. The observed density is within a few percentage points of the critical density, at the borderline between the two regimes. This is very peculiar and calls for an explanation.

The problem is that in the course of cosmic evolution the universe tends to be quickly driven away from the critical density. If, for example, we start 1 percent above the critical density at 1 second A.B., then in less than a minute we would get to twice the critical density and in a little over 3 minutes the universe would already have collapsed. Similarly, if we start 1 percent below the critical density, then in 1 year the density would be 300,000 times smaller than critical. In such a low-density universe, stars and galaxies would never form; there would be nothing but dilute, featureless gas. In order to have a nearly critical density at the present cosmic age of 14 billion years, the initial density has to be fine-tuned with a surgical accuracy. A calculation shows that at 1 second the density had to be equal to critical within 0.00000000000001 percent.

A closely related issue is the geometry of the universe. As we know from Friedmann, there is a connection between the density of the universe and its large-scale geometry. The universe is closed if the density is above critical, open if it is below critical, and flat if the density is exactly equal to critical. Thus, instead of asking why the density of the universe is so close to critical, we could just as well ask why its spatial geometry is so close to flat. That is why this fine-tuning puzzle is often called the *flatness problem*.

The horizon and flatness problems had been recognized since the 1960s, but had almost never been discussed—simply because no one had any idea as to what could be done about them. These problems could not be attacked without confronting an even greater puzzle that was looming behind them: What actually happened at the big bang? What was the nature of the force that caused the cosmic blast and sent all particles flying away from one another? With no progress in that direction for nearly half a century, physicists grew accustomed to the thought that this was one of those questions that you never ask—either because it does not belong to physics or because physics is not yet ready to tackle it. It therefore came as a total surprise when in 1980 Alan Guth made his dramatic breakthrough, pointing the way to resolve the stubborn cosmological puzzles in one shot.[1]

•

Guth came up with the idea that it was repulsive gravity that blew the universe up. He suggested that the early universe contained some very unusual material, which produced a strong repulsive gravitational force. If you ever try to give a talk with this kind of idea, you had better have a piece of antigravity stuff in your pocket, or at least be prepared to give a very good reason why anybody should believe that it really exists. Luckily for Guth, he did not have to invent any magic material. The leading elementary particle theories had it already in stock: it was called *false vacuum*.

FALSE VACUUM

"Can you make no use of nothing, nuncle?"
"Why, no, boy; nothing can be made out of nothing."
—SHAKESPEARE, *King Lear*

Vacuum is empty space. It is often regarded as synonymous with "nothing." That is why the idea of vacuum energy sounded so weird when Einstein first introduced it. But the physicist's view of the vacuum has been drastically transformed, as a result of developments in particle physics over the last three decades. The study of the vacuum still continues, and the more we learn about it, the more complex and fascinating it becomes.

According to modern theories of elementary particles, vacuum is a physical object; it can be charged with energy and can come in a variety of different states. In physics terminology, these states are referred to as different vacua. The types of elementary particles, their masses, and their interactions are determined by the underlying vacuum. The relation between particles and the vacuum is similar to the relation between sound waves and the material in which they propagate. The types of waves and the speed at which they travel vary in different materials.

We live in the lowest-energy vacuum, the *true vacuum*.[2] Physicists have accumulated a great deal of knowledge about the particles that inhabit this type of vacuum and the forces acting between them. The strong *nuclear force*, for example, binds protons and neutrons in atomic nuclei; the *electromagnetic force* holds electrons in their orbits around nuclei in atoms; and the *weak force* is responsible for the interactions of elusive light particles called neutrinos. As their names suggest, the three types of forces have very different strengths, with the electromagnetic force intermediate between the strong and the weak.

The properties of elementary particles in other vacua may be completely different. We do not know how many vacua there are, but particle physics suggests that apart from our true vacuum, there are likely to be at least two more, both having more symmetry and less diversity among particles and their interactions. The first is the *electroweak* vacuum, in which the electromagnetic and weak interactions have the same strength and are manifested as parts of a single, unified force. Electrons in this vacuum have zero mass and are indistinguishable from neutrinos. They dash about at the speed of light and cannot be captured into atoms. No wonder we do not live in this type of vacuum.

The second is the *grand-unified* vacuum, where all three types of particle interactions are unified. Neutrinos, electrons, and quarks (of which protons and neutrons are made) are all interchangeable in this highly symmetric state. While the electroweak vacuum almost certainly exists, the grand-unified vacuum is more speculative. Particle theories that predict its existence are attractive from the theoretical point of view, but they are concerned with extremely high energies, and observational evidence for these theories is scant and rather indirect.

Each cubic centimeter of the electroweak vacuum carries a huge energy and, by Einstein's mass-energy relation, a huge mass, approximately 10 million trillion tons (roughly the mass of the Moon). When faced with such colossal numbers, physicists resort to a shorthand power-of-ten notation. A trillion is 1 followed by 12 zeros; it is written as 10^{12}. Ten million trillion is 1 with 19 zeros; hence, the mass density of electroweak vacuum is 10^{19} tons per cubic centimeter. In a grand-unified vacuum, the mass density is still higher, by a whopping factor of 10^{48}. Needless to say, these vacua have never been synthesized in a laboratory: this would require energies far in excess of the present technological capabilities.

In contrast to these enormous energies, the energy of the normal, true vacuum is minuscule. For a long time it was thought to be exactly zero, but recent observations indicate that our vacuum has a small positive energy, which is equivalent to the mass of three hydrogen atoms per cubic meter. The significance of this finding will become clear in Chapters 9, 12, and 14.

High-energy vacua are called "false" because, unlike our true vacuum, they are unstable. After a brief period of time, typically a small fraction of a second, a false vacuum decays, turning into the true vacuum, and its excess energy is released in a fireball of elementary particles. We shall delve into the details of the vacuum decay process in the following chapter.

•

If vacuum has energy, then we know from Einstein that it should also have tension.[3] And, as we discussed in Chapter 2, tension has a repulsive gravitational effect. In the case of a vacuum, the repulsion is three times stronger than the attractive gravity caused by the mass, and the net effect is a strong repulsive force. Einstein used this antigravity of the vacuum to balance the gravitational pull of ordinary matter in his static model of the world. He found that the balance is achieved when the mass density of matter is twice that of the vacuum. Guth had a different plan: instead of balancing the universe, he wanted to blow it up. So he allowed the repulsive gravity of false vacuum to reign unopposed.

COSMIC INFLATION

What would happen if, at some early epoch, the space of the universe were in a false-vacuum state? If the matter density at that epoch were less than needed to balance the universe, then the repulsive gravity of the vacuum would have prevailed. This would cause the universe to expand—even if it did not expand to begin with.

To have a definite picture in mind, we shall assume that the universe is closed. Then it swells like an expanding balloon, as shown in Figure 3.1 in Chapter 3. As the volume of the universe grows, the matter is diluted and its mass density is reduced. But the mass density of false vacuum is a fixed constant; it always remains the same. Hence, very quickly the matter density becomes negligible, and we are left with a uniform, expanding sea of false vacuum.

The expansion is driven by the false-vacuum tension, overcoming the attractive force of the vacuum mass density. Since neither of these quantities changes with time, the *expansion rate* remains constant as well. The expansion rate tells by what fraction the universe grows in a chosen unit of time (say, in 1 second). Its meaning is very similar to that of the inflation rate in economics: the percentage increase of prices in a year. In 1980, when Guth

Figure 5.2. Alan Guth in his office at MIT. Guth is the proud winner of the 2005 contest for the messiest office, organized by *The Boston Globe.* (Photo by Larry Fink)

gave his seminar at Harvard, the rate of economic inflation in the United States was 14 percent. If it remained fixed at that value, prices would double every 5.3 years. Similarly, a constant expansion rate of the universe implies that there is a fixed time in which the size of the universe doubles.

The growth pattern with a constant doubling time is called exponential. It is known to build up very quickly to gigantic numbers. If a slice of pizza costs $1 now, then after 10 doubling cycles (53 years in our example) its price will be $1,024, and after 330 cycles it will be 10^{100}. This stupendous number, 1 followed by 100 zeros, has a special name: a *google*. Guth suggested that we adopt the term "inflation" in cosmology, using it to describe an exponential expansion of the universe.

The doubling time in a false-vacuum universe is unbelievably short. The higher the vacuum energy, the shorter the time. For the electroweak vacuum, the universe would expand by a google in one-thirtieth of a microsecond, and for the grand-unified vacuum this expansion would happen 10^{26} times faster. In this tiny fraction of a second, a region the size of an atom would be blown to dimensions much greater than the entire currently observable universe.

Since false vacuum is unstable, it eventually decays, and its energy ignites a hot fireball of particles. This event signals the end of inflation and the starting point of the usual cosmological evolution. We thus get an enormous, hot, expanding universe from a tiny initial seed. As an extra bonus, amazingly, the horizon and flatness problems of big bang cosmology disappear in this scenario.

The essence of the horizon problem is that the distances between some parts of the observable universe appear to have always been greater than the distance traveled by light since the big bang. This implies that these regions have never interacted with one another, and then it is hard to explain how they could reach nearly identical temperatures and densities. In the standard big bang theory, the distance traveled by light grows proportionally to the age of the universe, while the separation between the regions increases more slowly, because cosmic expansion is being slowed down by gravity. Regions that cannot interact now will be able to do so in the future, when the light-travel distance finally catches up with their separation. But at earlier times the light-travel distance fell even shorter of the mark, so that if the regions cannot interact at the present epoch, they were surely unable to do so in the past. The root of the problem can thus be traced to the attractive nature of gravity, which causes the expansion to slow down with time.

In a false-vacuum universe, however, gravity is repulsive; so instead of slowing down, expansion accelerates. Then the situation is reversed: regions that can exchange light signals will lose their ability to interact in the future. And more important, regions that are out of each other's reach must have interacted in the past. The horizon problem has disappeared!

The flatness problem dissolves just as easily. It turns out that the universe is driven away from the critical density only if its expansion is slowing down. In the case of accelerated, inflationary expansion, the opposite is true: the universe is driven *toward* the critical density, and thus to flatness. Since inflation enlarges the universe by an enormous factor, we can see only a tiny part of it. This observable region appears to be flat, just as the surface of the Earth appears flat when you look at it from close by.

In summary, a brief period of inflation makes the universe large, hot, uniform, and flat, setting just the right initial conditions for the standard big bang cosmology.

The theory of inflation was about to begin its conquest of the world. As for Guth himself, his days as a postdoc were over. He accepted a job offer from his alma mater, the Massachusetts Institute of Technology, where he has remained ever since.

This would be a nice happy ending to the story of inflation, except for one unfortunate problem: the theory did not work.

6

Too Good to Be Wrong

Truth comes out of error more readily than out of confusion.
—FRANCIS BACON

THE GRACEFUL EXIT PROBLEM

Every physicist knows the sinking sensation of discovering a fatal flaw in the beautiful theory that you thought up a few days ago. Alas, this is the fate of most beautiful theories. And so it was with inflation. As usual, the devil was in the details. On a closer examination, the false vacuum did not decay as smoothly as anticipated.

The process of vacuum decay is similar to the boiling of water. Small bubbles of true vacuum pop out randomly and expand in the midst of false vacuum (Figure 6.1). As the bubbles grow, their interiors remain nearly empty and all the energy released from converting false vacuum into true is concentrated in the expanding bubble walls. When bubbles collide and merge, their walls disintegrate into elementary particles. The end result is true vacuum filled with a hot fireball of matter.

This is indeed what happens if bubbles pop out at a feverish rate, so that the whole decay process is complete in less than one doubling time. That would mean, however, that inflation ends too soon, way before the universe becomes homogeneous and flat. We are interested in the opposite case,

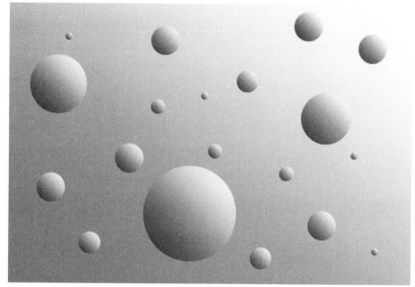

Figure 6.1. Small bubbles of true vacuum pop out randomly and expand. Bubbles that formed earlier have grown to a bigger size.

when the bubble-formation rate is low, so that the universe can expand by a large factor before bubbles begin to collide. But, as the Swiss physicist Paul Ehrenfest used to say, that's where the frog jumps into the water.

The trouble is that the space between the bubbles is filled with false vacuum and is therefore rapidly expanding. The bubbles grow very fast, at speeds approaching the speed of light, but even that is no match for the exponential expansion of false vacuum. If the bubbles do not collide within one doubling time after they are formed, then at later times their separation will only grow, so they will never collide.

The conclusion is that inflation cannot possibly end. The bubbles grow to unlimited sizes, and new, small bubbles keep popping out in the expanding gaps between them. As a result, the wonderful uniformity created by inflation is completely destroyed. The absence of a suitable ending to inflationary expansion has become known as the *graceful exit problem*.

Guth realized that there was a problem a few months after he went public with his new theory. His paper on inflation was not yet written at that time, and for a very simple reason: Alan Guth is the biggest procrastinator in the world. (I learned this firsthand after working with him on a number of research projects.) Of course, Guth was disappointed to find a serious flaw in his theory. But still he felt the idea was too good to be wrong. When he finally got around to writing the paper in August of 1980, he concluded it with these words: "I am publishing this paper in the hope that it will . . . encourage others to find some way to avoid the undesirable features of the inflationary scenario."[1]

SCALAR FIELD

To get to the root of the problem, let us now discuss the false-vacuum decay in more detail. The decay process was studied by the Harvard physicist Sidney Coleman, who described it in terms of *scalar fields*.

A *field* is a quantity that has some value at every point in space. The values may vary from one point to another and can also change with time. A simple example of a field is the temperature. The North Pole, the tip of Cape Cod, the center of the Sun—all points in the universe have a certain value of temperature. Another familiar example is the magnetic field. In addition to its magnitude, this field also has a direction. We don't feel the magnetic field, but its presence becomes evident when we examine a compass. The compass needle will point in the direction of the field, and the field strength can be judged by how forcefully it causes the needle to swing in that direction.

Fields like the temperature, which do not have any direction, are called scalar fields. They are characterized by a single number: their magnitude. Scalar fields play an important role in elementary particle physics. According to modern particle theories, the space of the universe is pervaded by a number of scalar fields, whose values determine the vacuum energy, as well as the particle masses and their interactions. In other words, these fields determine what vacuum we live in. At present, the scalar fields are at their true vacuum values, but things could have been different at earlier epochs.

To illustrate the physics of vacuum decay, we shall consider a single scalar field and focus on how it affects the vacuum energy. Each cubic centimeter of space has energy, which depends on the magnitude of the field. The exact dependence is not currently known, but its general form is expected to resemble a hilly landscape, as in Figure 6.2, with hilltops (maxima) at some values of the field and valleys (minima) at others. The behavior of the scalar field is very similar to that of a ball rolling along the terrain depicted in this energy landscape. The value of the field is represented by the location of the ball along the horizontal axis. Depending on the initial position of the ball, it will roll down to one or the other energy minimum in the figure. The lower minimum has almost zero energy density; it corresponds to the true vacuum. The upper minimum corresponds to a high-energy false vacuum.

Suppose now that we start with a false vacuum at every point in space. This situation is represented by the ball lying in the upper minimum. It will lie there for a very long time, unless someone kicks it upward, supplying the energy needed to go over the barrier and into the lower minimum. But according to the quantum theory, objects can "tunnel" through energy barriers. If you were to observe such an event, you would see the ball disappear and instantly materialize on the other side of the barrier.

Quantum tunneling is a probabilistic process. You cannot predict exactly when it will happen, but you can calculate the probability for it to happen in a given interval of time. For a macroscopic object, like a ball, the tunneling probability is extremely low. If, for example, you want a can of

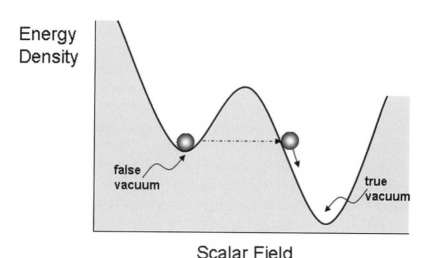

Figure 6.2. The energy landscape of a scalar field with a false and a true vacuum. The field can tunnel through the barrier separating the two vacua.

Coke to tunnel out of a vending machine, you will have to wait for much longer than the present age of the universe. But in the microscopic world of elementary particles, quantum tunneling is much more common. As I mentioned in Chapter 4, George Gamow used the tunneling effect to explain the decay of radioactive atomic nuclei. In the case of a false vacuum, the probability that a large region of space will tunnel to the true vacuum is completely negligible. The tunneling occurs in a tiny, microscopic region, resulting in a small true-vacuum bubble. This is the bubble-formation process that we discussed in the preceding section of this chapter. The tunneling probability may be large or small, depending on the shape of the energy function. (The probability is large for low and narrow energy barriers.)

Despite the similarity between the tunneling of a ball and that of a scalar field, there is also an important difference. The ball tunnels between two different points in space, while for a scalar field the tunneling occurs between two different values of the field at the same location.

What has transpired from this analysis is that if there is an energy barrier between the two vacua, then false-vacuum decay can proceed only through quantum tunneling. The tunneling results in a haphazard pattern of bubbles that never merge, so the decay process is never complete. But what would happen if we remove the barrier?

SLOW DOES IT

Andrei Linde, a young Russian cosmologist, was the first to consider unorthodox scalar field models that had no barrier between the false and true vacua.

As before, suppose we start with a small closed universe and a scalar field in the false-vacuum state. If there is no barrier, the ball representing the field simply rolls down toward the true vacuum (see Figure 6.3). There are no bubbles, and the field remains uniform in the entire space as it rolls downhill. When it gets to the bottom, the scalar field starts oscillating back and forth. The energy of the oscillations is then quickly dissipated into a fireball of particles, while the field settles at the energy minimum.

The problem is, however, that in the absence of a barrier the field would roll down very fast and inflation would be cut off too early. Having recognized this danger, Linde made a crucial step. He suggested that the energy function should have the form of a hill with a very gentle slope, as shown in Figure 6.4. The flat region near the hilltop in the figure plays the role of the false vacuum. If the ball is placed somewhere in that region, it will start rolling *very* slowly. And since the slope is so flat, the ball will remain at about

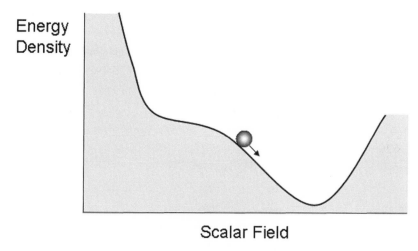

Figure 6.3. The energy landscape without a barrier. The scalar field quickly rolls down to the true vacuum.

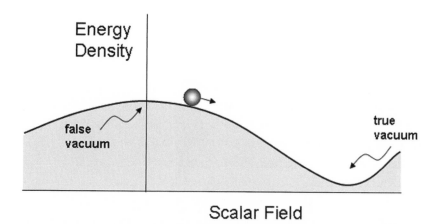

Figure 6.4. A "flattened hill" energy landscape. The scalar field slowly rolls downhill, while inflation continues.

the same height. Now, remember, the height in this figure stands for the energy density of the scalar field, and keeping it constant is all that is needed to sustain the constant rate of inflation.

Linde's key observation was that in the flat region near the hilltop, the scalar field rolls very slowly, and therefore it takes a long time to traverse that region. In the meantime, the universe expands exponentially, resulting in a huge expansion factor. When the field gets to the steeper part of the energy slope, it rolls down faster, and when it finally reaches the minimum, it oscillates and dumps its energy into a hot fireball of particles. At this point we have an enormous, hot, expanding universe, which is also homogeneous and nearly flat. The graceful exit problem has been solved!

All that is needed is a scalar field whose energy function has the form of a flattened hill, as in Figure 6.4. You may be wondering how the scalar

Figure 6.5. Andrei Linde (left) with Slava Mukhanov (of Ludwig-Maximilians University in Munich). (Courtesy of Sugumi Kanno)

field got on the top of the hill to begin with. Good question. But wait until Chapter 17.

Linde's paper appeared in February of 1982, and a few months later essentially the same idea was published independently by the American physicists Andreas Albrecht and Paul Steinhardt. The theory of inflation had been rescued.

Another important question is whether or not such scalar fields really exist in nature. Unfortunately, we don't know. There is no direct evidence for their existence. Scalar fields appearing in the simplest electroweak and grand-unified theories have energy functions that are too steep for the purposes of inflation. But there is a class of *supersymmetric* theories, which include plenty of scalar fields with flat energy functions. The *superstring theory*, which is now the leading candidate for the fundamental theory of nature, belongs to this class. (We shall have more to say about superstrings in Chapter 15.)

THE NUFFIELD WORKSHOP

The next act of the drama was set in the medieval university town of Cambridge. By invitation of Stephen Hawking, some thirty cosmologists from all over the world converged there in the summer of 1982. They gathered for a three-week workshop on the very early universe, funded by the Nuffield Foundation. I was thrilled to be among the participants: Hawking asked me to talk about my recent work on cosmic strings.

I instantly fell in love with Cambridge. Early in the morning I would get up to stroll through the grounds of old colleges. Gothic chapels, clock towers, austere walled courts with their perfect rectangular lawns and bright splashes of flowers—vestiges of another, more contemplative age. By nine o'clock I would be back to modernity, sitting in the conference hall and waiting for the talks to begin. Thankfully, there were only two talks a day, one in the morning and one in the afternoon, with plenty of time left for informal discussions. British food was not among the highlights of the trip, but British beer is quite a different story, and I spent many evening hours discussing physics and other matters over a pint of lager.

The program of the meeting emphasized recent developments in cosmology, and inevitably the theory of inflation took the center stage. The graceful exit problem was now out of the way, but still there was another major concern.

It is true that inflation makes the universe flat and smooth, but perhaps it does too good a job of that. No galaxies or stars would ever form in a perfectly homogeneous universe. As we discussed in Chapter 4, galaxies have evolved from small variations in the density. The origin of these primordial inhomogeneities, or density perturbations, became the central issue of the workshop.

Shortly before the meeting, Hawking wrote a paper with a very interesting idea. According to the quantum theory, the evolution of all physical systems is not entirely deterministic, but is subject to unpredictable quantum jerks. So, as the scalar field rolls downhill, it experiences random kicks back and forth. The directions of the kicks are not the same in different regions of the universe, and as a result, the scalar field arrives at the bottom of the hill at slightly different times in different places. In regions where inflation lasted a little longer, the matter density would be slightly higher.* Hawking's idea was that the resulting small inhomogeneities led to the formation of galaxies and galaxy clusters. If he was right, then quantum effects, which are normally important on tiny, subatomic scales, were responsible for the existence of the largest structures in the universe!

Naturally, Guth was very excited by this development. Not only did it resolve the difficulty of the theory, but also it opened the tantalizing possibility of testing inflation observationally. Density perturbations can be observed through cosmic microwaves and then compared with the predictions of the theory. This was tremendously important!

The calculation of density inhomogeneities produced during inflation is a very challenging technical problem. Hawking's paper gave very few details and was difficult to follow. So Guth joined forces with a Korean-born physicist, So-Young Pi, to work out the perturbations using a method that they could both understand. They were not quite done when Guth had to leave for the Nuffield Workshop, and he finished the calculation during his first days in Cambridge. To his great surprise, the result was very different from Hawking's. They both found that the perturbations depended on the

---

*After the end of inflation, the matter density is diluted by the expansion of the universe. Therefore, regions of space that were in a hurry to end inflation are already diluted by the time other sluggish regions finally end inflation.

form of the scalar field energy landscape. But the dependence was different, and Guth's answer gave a much larger magnitude for the perturbations.

Guth discussed the matter with Hawking, but the difference remained unresolved. Hawking insisted on his result. When Guth told me about their conversation over lunch, he looked puzzled. He was not sure his answer was correct and said he would have to recheck several points in the calculation.

To add to the confusion, there was yet another group working on the same problem. Paul Steinhardt had calculated the inhomogeneities in collaboration with two other American cosmologists, Jim Bardeen and Michael Turner. They also disagreed with Hawking, but their answer was much smaller! Finally, there was a Russian physicist, Alexei Starobinsky, who was also scheduled to talk on the subject of density perturbations. But he kept to himself, and nobody knew what result he was going to announce.

Starobinsky was not a novice to cosmology. Among other things, he was known for inventing a version of inflation about a year earlier than Guth. The rub was that he invented it for the wrong reason. He thought his model could remove the initial singularity—which it could not. But he did not realize that it could solve the horizon and flatness problems. Without this crucial insight, the model did not get much notice at the time, but now it is regarded as a viable alternative to the scalar field models of Linde, Albrecht, and Steinhardt.[2]

Starobinsky was scheduled to speak first. His style of presentation was typical of the Russian school of physics and could be traced back to one of its originators, the Nobel Prize laureate Lev Landau. At Landau's famous weekly seminar, the speaker was presumed to be an idiot and had a narrow window of opportunity to prove otherwise at the beginning of the talk. So the seminars were given mainly "for Landau," to convince him that the speaker knew what he was talking about, and without undue concern that the talk might go above the heads of almost everybody else. Now, add to this a Russian accent and a strong stutter, and you will not be surprised that Starobinsky's talk was not easy to follow. Yet, by the time he was finished, one thing was clear: he had found the inhomogeneities to be large, pretty close to Guth's result.

The next day it was Hawking's turn to speak. The legendary physicist suffers from Lou Gehrig's disease and has been wheelchair-bound since the early 1970s. He now communicates through a voice synthesizer, selecting words one by one from a menu on a computer screen. At the time of the meeting he could still speak, but barely. Most people could not understand him, and one of his students served as an interpreter during the talk. Hawking's lecture followed the line of argument in his paper, but at the end there was a surprise. The last step of the calculation was now different, and the result was the same as found by Guth and Starobinsky! After talking to Guth and hearing Starobinsky's lecture, Hawking must have spotted an error in his calculation. He never mentioned, though, that he was correcting an error in his paper, or that his new result was also derived by Starobinsky and Guth.

The majority of the talks at the Nuffield Workshop were on the subject of inflation, and despite much excitement about the new theory, it was a bit of an overdose. The talks on other early-universe topics provided a welcome relief—the sentiment I tried to express in the opening slide of my lecture on cosmic strings (Figure 6.6). Strings are line-like relics of the hot, high-energy epoch in the early universe. They are thin tubes of false vacuum, which are predicted in some particle physics models. In my talk I discussed the formation of strings and their possible astrophysical effects. The talk was well received, and I could now sit back, relax, and watch the final stretch of the race to figure out the density perturbations.

Steinhardt and his friends were still holding out. They were concerned about some subtle points in their calculation and kept working furiously to

Figure 6.6. Inflation overdose—the opening slide of my talk on cosmic strings.

clear them up. The answer they were getting was still much smaller than Hawking's original result.

Guth was scheduled to speak during the third week of the meeting. He worried that Steinhardt and company might give him a hard time and used every opportunity to retreat into his room and check various parts of his calculation. He later realized that he had even missed the conference banquet while preparing for his talk.

Despite mounting tension, the battle was not to happen. A few days before the talk, Steinhardt and his collaborators conceded defeat. They found some errors in the approximations that they had used, and now their result was in agreement with the other contestants'. Guth's talk went very smoothly: he reiterated the original result that he had obtained earlier. Thus, by the end of the workshop, all four participating teams had reached a full consensus.

The final surprise of this remarkable race came long after the workshop was over. Much to their dismay, the former contestants discovered that the problem of quantum-induced density perturbations that they worked so hard to untangle had already been solved—a full year before they crossed swords in Cambridge. The solution was published by two Russian physicists, Slava Mukhanov* and Gennady Chibisov, from the Lebedev Institute in Moscow.[3] They worked out perturbations for the Starobinsky version of inflation, but the calculation was essentially the same as for the scalar field models. You can often find something interesting by reading Russian physics journals!

•

The end point of the calculations was a formula for the magnitude of density perturbations produced by quantum jitters of the scalar field as it rolls downhill during inflation. This magnitude depends on the shape of the energy landscape and also on the size of the region where the perturbation occurs. Cosmic structures span a wide range of distance scales. The scale of stars is much smaller than that of galaxies, which is in turn smaller than the scale of galaxy clusters. The magnitude of perturbations on these vastly different scales could well be very different. But the formula says that all per-

*Mukhanov is now at the Ludwig-Maximilians University in Munich; see his photo on p. 60.

turbations are created very nearly equal. From the smallest cosmic structures to the largest, their magnitude changes by no more than 30 percent.

This property of scale-independence of the inflationary perturbations is not difficult to understand. The quantum kicks initially affect the scalar field in a tiny region of space, but then the perturbation is stretched to a much greater size by the exponential expansion of the universe. Perturbations produced earlier during inflation are stretched for a longer time and encompass a larger region. But the magnitude of the perturbation is set by the initial quantum kick, which is pretty much the same for all relevant scales.*

Scale-independence of the density perturbations can be used to derive predictions for variations in the intensity of cosmic microwaves over the sky and, ultimately, to test inflation. A speculative hypothesis about the early moments of the universe has thus been transformed into a testable physical theory. But it took another decade before the theory of inflation was put to the test.

A RECIPE FOR OVERNIGHT SUCCESS

It usually takes years, if not decades, for a new theory to be widely accepted. Physicists may appreciate a beautiful idea, but they will only be convinced when predictions of the theory are confirmed by experiments or by astronomical observations. This is particularly true in cosmology, where observers have always had a hard time keeping up with the imagination of the theorists, and the big bang theory is as good an example as any. The papers by Alexander Friedmann remained unnoticed until after his death, and the work of George Gamow was all but ignored for more than a decade. What a contrast to how inflation was received!

Nearly forty papers were published on the new theory in the first year after Guth's original paper. In a couple of years, this number climbed to two hundred and remained more or less steady at about two hundred papers a year for the following decade. It looked as if people dropped whatever they were doing and started working on the theory of inflation.

Why was inflation such an instant success? In part, this was due to soci-

---

*As the scalar field slowly rolls down the energy slope, the kicks get weaker and the resulting perturbations smaller. But the downhill roll of the field is so slow that it does not move much during the time that it generates perturbations on all astrophysically relevant scales.

ological reasons. Particle physicists had just finished developing theories of strong and electroweak interactions. There was a small army of them, and suddenly they found themselves with little to do. New ideas in particle physics were all related to extremely high energies. There was no way to test these theories in the existing particle accelerators, so progress had stalled. The only accelerator that could boost particles to the required energies appeared to be the big bang, and particle physicists were increasingly turning their sights to cosmology as a testing ground for new ideas. By the early 1980s, a mass conversion was under way from particle physics to cosmology. The converts were new to the field and were looking for interesting problems to solve.

It was on this background that Guth suggested his idea of inflation. He gave physicists exactly what they were looking for. It really helped that Guth's theory was incomplete. If you fully solve an important problem, your work may be admired, but you do not create an industry. Inflation, on the other hand, was just an outline of a theory, with many blanks to be filled. It offered plenty of problems to work on and to give to your graduate students.

But, apart from sociology, the long-term popularity of inflation is due to the appeal and the power of the idea itself. In some ways, inflation is similar to Darwin's theory of evolution. Both theories proposed an explanation for something that was previously believed to be impossible to explain. The realm of scientific inquiry was thus substantially expanded. In both cases, the explanation was very compelling, and no plausible alternatives have ever been suggested.

Another parallel with Darwin is that the idea of inflation was already in the air at the time when Guth proposed it.* Guth's key contribution was that he clearly realized what inflation was good for, providing the motivation to solve the graceful exit and other problems of inflation.

UNIVERSE AS A FREE LUNCH

We have assumed so far that the starting point for inflation was a small closed universe with a scalar field in the false vacuum, at the top of its energy hill.

*Erast Gliner, Starobinsky, and Linde in Russia; Katsuhiko Sato in Japan; and Robert Brout, François Englert, and Edgard Gunzig in Belgium were all considering a possible period of exponential expansion in the early universe. Sato was also aware of the graceful exit problem.

But these assumptions are not necessary. We could instead have started with a small chunk of false vacuum in an infinite universe. Such a beginning would also lead to inflation, but in a somewhat unexpected way.

Remember, false vacuum has a large tension, which is responsible for its repulsive gravity. If it fills the entire space, the tension is the same everywhere and has no physical effect other than gravitational. But if it is surrounded by true vacuum, the tension inside is not balanced by any force outside and causes the false vacuum chunk to shrink. You might think that tension would be counteracted by the repulsive gravity, but this is not what actually happens.

Analysis based on Einstein's general relativity shows that the gravitational repulsion is purely internal. So, if you had a false-vacuum chunk for your lecture demonstration, objects would not fly away from it as in Figure 1.1. They would be attracted to it instead. Outside the false vacuum, the gravitational force is attractive as usual. So the force of tension causes the chunk to shrink, while its interior "wants" to expand because of the internal gravitational repulsion. The outcome depends on the size of the chunk.

If it is smaller than a certain critical size, the tension wins, and the chunk shrinks like a piece of stretched rubber. Then, after a few oscillations, it disintegrates into elementary particles.

If the size is bigger than critical, repulsive gravity wins and the false vacuum begins to swell. As it does so, it warps space, like a blown-up balloon. This effect is illustrated in Figure 6.7 for a spherical false-vacuum region. Only two spatial dimensions are shown, so the spherical boundary of the region is represented by a circle. Tension pulls the boundary inward, toward the center of the sphere, and this has the effect of reducing the volume of false vacuum. But this reduction is totally negligible compared with the exponential expansion of the interior.

The inflating balloon is connected to the exterior space by a narrow "wormhole." From outside, the wormhole is seen as a black hole, and observers in the exterior region can neither verify nor disprove that there is a huge inflating universe inside this black hole. Likewise, observers that will evolve in the inflating bubble universe will see only a tiny part of it and will never find out that their universe has a boundary and that there is another big universe beyond it.

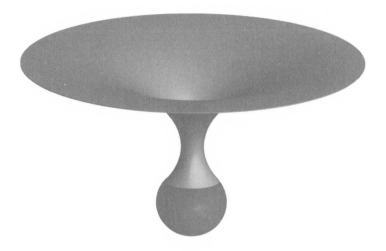

Figure 6.7. An inflating false-vacuum balloon (dark) is connected to the exterior space by a "wormhole" and is seen as a black hole from the exterior region.

Since the fate of the false-vacuum sphere depends so crucially on whether its radius is greater than critical, it is important to know what the critical radius actually is. The answer depends on the vacuum energy density: the larger the energy density, the smaller the critical radius. For the electroweak vacuum it turns out to be about 1 millimeter, and for the grand-unified vacuum it is 10 trillion times smaller. This is all that is needed to create a universe! Truly, the ultimate free lunch. Almost . . .

PART II

. . .

ETERNAL INFLATION

7

The Antigravity Stone

It would be more impressive if it flowed the other way.
—OSCAR WILDE, on Niagara Falls

The theory of inflation became a major topic of my research soon after that Wednesday seminar at Harvard in 1980, where I first heard about it. In fact, if I were more mystically inclined, I might have seen the writing on the wall even before Guth's seminar. There were some clues pointing to repulsive gravity right where I work, at Tufts University.

Set on a gently sloping hill, amid shady elms, the Tufts campus exudes an air of grace and tranquility. As you climb the stairs up the hill to the heart of the campus, and walk past the ivy-covered Romanesque chapel, you may notice a peculiar monument. It is a sizable slab of granite, rising vertically from the ground, like an old tombstone. The inscription says:

> THIS MONUMENT HAS BEEN
> ERECTED BY THE
> GRAVITY RESEARCH FOUNDATION,
> ROGER W. BABSON FOUNDER.
> IT IS TO REMIND STUDENTS OF
> THE BLESSINGS FORTHCOMING
> WHEN A SEMI-INSULATOR IS

DISCOVERED IN ORDER TO HARNESS

GRAVITY AS FREE POWER

AND REDUCE AIRPLANE ACCIDENTS.

1961

This is the notorious antigravity stone, the sign of my destiny.

Roger Babson, who also founded Babson College, was living proof that shrewd business judgment can peacefully coexist with far-out scientific ideas. He claimed it was by using Newton's laws of mechanics that he predicted the stock market crash of 1929 and the Great Depression that followed. With Newton's help, he managed to amass a great fortune, and in gratitude to Sir Isaac, he bought an entire room from Newton's last residence in London as well as an apple tree that is a descendant of the famous tree at Newton's family home in Lincolnshire. Legend has it that the fall of an apple from that tree inspired Newton to discover the law of gravity. And gravity, as you might have guessed, was a paramount theme in Babson's universe.

Babson's obsession with gravity dates back to his childhood, when his sister drowned in a river. He blamed gravity for her death and resolved to free humanity from its fatal pull. In his book *Gravity—Our Enemy No. 1*, Babson described the benefits to be derived from an insulator against gravity. It would reduce the weight of airplanes and increase their speed; it could even be used in the soles of shoes to lighten weight when walking. Babson's lifelong friend, the famous inventor Thomas Edison, suggested to him that birds may have some antigravity stuff in their skin, and Babson promptly acquired a collection of some five thousand stuffed birds. It is not clear exactly what he did with them, but apparently this line of research did not result in any breakthrough.

To his credit, Babson did put his money where his mouth was. He made gifts to several universities, Tufts included, to facilitate antigravity research. The only condition of the grant was that a monument with Babson's inscription be erected on campus.

The wacky monument was a source of embarrassment for the Tufts administration and inspired numerous pranks by the students. It would occasionally disappear, only to reappear where it was least expected. Once it was found blocking the entrance of the trustees and the president at the commencement. At one time it looked as if the stone had disappeared for good, but then it miraculously resurfaced ten years later. It turned out that a group of students had buried it somewhere on campus and then dug it out when they returned to Tufts for their class reunion. Gravity alone was clearly not enough to keep the stone in place, so it was finally cemented to the ground.

Since few scientists could claim they had an active program of research in antigravity, the Babson money proved rather difficult to get. It's not that nobody tried: the university president, Jean Mayer, who was a nutritionist, argued unsuccessfully that weight loss was antigravity. After years of discussions and legal arguments, the money was eventually used to establish the Tufts Institute of Cosmology.

Like any self-respecting academic institution, our institute has its own unique ritual—an "inauguration" ceremony for cosmology Ph.D. recipients. After defending the dissertation, a new Ph.D. gets an apple dropped on his

Figure 7.1. A triumphant Dr. Vitaly Vanchurin after his Ph.D. inauguration, surrounded by members of the Institute of Cosmology. Standing, from left to right: Larry Ford, Ken Olum, and the author. (Courtesy of Delia Schwartz-Perlov)

head while kneeling in front of the antigravity stone. The apple comes from the hand of the thesis advisor and may then be eaten by the "inauguree."

By the time the Institute of Cosmology was established, Babson was long since dead and his Gravity Research Foundation had evolved into a respectable institution giving research grants on gravitation. Nobody really expected that Tufts cosmologists would work on antigravity, but strangely enough—they do. Much of the research at the institute is focused on false vacuum and its repulsive gravity, which certainly qualifies as antigravity. So I think Mr. Babson could not have found a better use for the money. We have not succeeded in reducing the number of airplane accidents though.

8

Runaway Inflation

*In my opinion, the most plausible answer to what happened be-
fore inflation is—more inflation.* —ALAN GUTH

UNIVERSE BEYOND THE HORIZON

What lies beyond our present horizon? That was the question that intrigued me from the early days of inflation. If we can see only a minuscule part of the universe, then what is the big picture— like the view of our planet that is revealed to space travelers as their spaceship leaves the Earth?

The theory of density perturbations provided some clues. According to this theory, the pattern of how galaxies are distributed in space is determined by quantum kicks experienced by the scalar field during inflation. This is a random process; so some regions of the same size as ours have more galaxies, and others have less. The reason we have the Milky Way galaxy right here, where we are, is that the scalar field at this location had a tiny kick backward, away from the true vacuum, so that it ended its roll down the energy hill a bit later than it did in the neighboring locales. This produced a small density enhancement, which later evolved into our galaxy. Similar little humps on the smooth density background gave rise to our neighbor Andromeda and to countless other galaxies within our horizon and beyond. This description of structure formation suggests that the remotest parts of the universe look more or less the same as what we see around here. But I was beginning to suspect that something was missing from this picture.

The effect of quantum kicks is very small because they are much weaker than the force due to the slope of the energy hill that drives the scalar field down. This explains why the field reaches the bottom everywhere at about the same time and only small density perturbations are produced. The question I was asking myself was, What happens when the field is near the top of the hill, where the slope is very small? There, it should be at the mercy of quantum kicks, which shove it at random one way and then the other. The universe resulting from inflation might then be far less orderly, and more erratic, than it first appeared.

To picture the behavior of the scalar field near the top of the hill, we shall use a politically incorrect but pertinent analogy. Let me introduce a gentleman, named Mr. Field, who had too much to drink and is now trying to maintain his vertical position. He has little control of his feet and no idea where he is heading, so he steps to the right or to the left completely at random. Mr. Field starts his promenade at the top of the hill, as in Figure 8.1. Since on average he steps to the right as frequently as he does to the left, he is not getting anywhere very fast. But after a large number of steps he will gradually drift away from the hilltop. Eventually, he will get to the steeper

Figure 8.1. Mr. Field walks randomly on the flat portion of the hill and slides down when he gets to the steeper slope.

part of the hill, where he will inevitably slip and finish the rest of the journey sliding downhill on his rear.

The scalar field behavior during inflation is very similar. The field wanders aimlessly near the top of the energy hill, until it reaches a steeper slope; then it "rolls" down toward the end of inflation. On the flat portion near the top of the hill, the field variation is induced by quantum kicks and is totally random, while the roll down the slope is orderly and predictable, with quantum kicks introducing only a small disturbance. The time interval between successive quantum kicks is roughly equal to the doubling time of inflation. This means that Mr. Field takes about one step per doubling time. And since he makes many steps wandering around the flat hilltop, it follows that the false vacuum takes many doubling times to decay.

A particular sequence of steps that brings Mr. Field from the top of the hill to the bottom represents one possible history of the scalar field. But quantum kicks experienced by the field differ from one place to another, so the scalar field histories will also be different. Each quantum kick affects a small region of space. Its size is, roughly, the distance traveled by light during one doubling time of inflation; we shall call it a "kickspan."* We can imagine a party of gentlemen in the same condition as Mr. Field, each representing the scalar field at some point in space. When two points are within a kickspan of one another, they experience the same quantum kicks; so the corresponding two gentlemen do all the steps in sync, like a pair of tap dancers. But the points are rapidly driven apart by the inflationary expansion of the universe, and when they get separated by more than a kickspan, the two gentlemen part company and start walking independently. Once this happens, the scalar field values at the two points start gradually drifting apart, while the distance between the points continues to be rapidly stretched by inflation.

The smallness of density perturbations in our observable region tells us that all points in this region were still within a kickspan of one another when the scalar field was well on its way down the hill. That is why the effect of quantum kicks was very minor, and the field reached the bottom everywhere at about the same time. But if we could go to very large distances, far

---

*This is the maximum distance over which communication is possible in the inflating universe. It is the same as the critical size necessary for a chunk of false vacuum to inflate (see Chapter 6): 1 millimeter for electroweak vacuum and 10^{13} times smaller for grand-unified vacuum. This distance plays the role of the horizon in the inflating universe; I use a different term—"kickspan"— to avoid confusion with the present horizon.

beyond our horizon, we would see regions that parted our company when the field was still wandering near the hilltop. The scalar field histories in such regions could be very different from ours, and I wanted to know what the universe looks like on such superlarge scales.

Imagine a large party of jolly walkers starting off from the top of the hill. Each walker represents a remote region of the universe, so they all walk independently. If the flat portion of the hill is N steps long, then a typical random walker will cross it in about N^2 steps. Roughly half of the walkers will do it faster and another half slower. For example, if the distance is 10 steps, it will take, on average, 100 steps for a random walker to cross it. So, after the first 100 steps, about half of the company will have reached their destination at the bottom, while the other half will still be enjoying the promenade. In another 100 steps, the number of remaining walkers will again be cut in half, and so on, until the last fellow finally slides downhill.

But now, there is a crucial difference between the walkers and the inflating regions they represent. While a walker is wandering near the hilltop, the corresponding region undergoes exponential inflationary expansion. So the number of independently evolving regions rapidly multiplies. This is as if the walkers were to multiply as well. As I continued to think about this, the picture was gradually taking shape.

ETERNAL INFLATION

In a way, inflation is similar to the reproduction of bacteria. There are two competing processes: bacteria reproduce by division, and they are occasionally destroyed by antibodies. The outcome depends on which process is more efficient. If the bacteria are destroyed faster than they reproduce, they will quickly die out. Alternatively, if the reproduction is faster, bacteria will rapidly multiply (Figure 8.2).

In the case of inflation, the two rival processes are the false-vacuum decay and its "reproduction" due to the rapid expansion of inflating regions. The efficiency of decay can be characterized by a *half-life**—the time during which half of the false vacuum would decay if it were not expanding. (In

---

*The term "half-life" is borrowed from nuclear physics, where it refers to the time during which half of the atoms in a sample of radioactive material will decay.

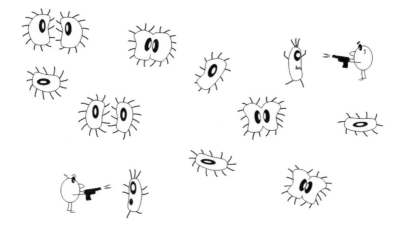

Figure 8.2. The number of bacteria will rapidly grow if the bacteria reproduce faster than they are destroyed.

our random-walk analogy, this is the time it takes for the number of walkers to be reduced by half.) On the other hand, the efficiency of reproduction is characterized by the doubling time, during which the volume of inflating false-vacuum doubles. The false-vacuum volume will shrink if the half-life is shorter than the doubling time and will grow otherwise.

But it follows from the discussion in the preceding section that the false-vacuum half-life is long compared with the doubling time. The reason is that in models of inflation the energy hill is rather flat and it takes many steps to cross it. Since each step of a random walker corresponds to one doubling time of inflation, it follows that the half-life must be much longer than the doubling time. Hence, false-vacuum regions multiply much faster than they decay. This means that inflation never ends in the entire universe and the volume of inflating regions keeps growing without bound!

At this very moment, some distant parts of the universe are filled with false vacuum and are undergoing exponential inflationary expansion. Regions like ours, where inflation has ended, are also constantly being produced. They form "island universes" in the inflating sea.* Because of inflation, the space between these islands rapidly expands, making room for more island

*Guth calls these islands "pocket universes." But, as Leonard Susskind has noted, this tends to ruin the prose.

universes to form. Thus, inflation is a runaway process, which stopped in our neighborhood, but still continues in other parts of the universe, causing them to expand at a furious rate and constantly spawning new island universes like ours.

The energy of the decaying false vacuum ignites a hot fireball of elementary particles and triggers the formation of helium and all the subsequent events of the standard big bang cosmology. Thus, the end of inflation plays the role of the big bang in this scenario. If we make this identification, then we should not think of the big bang as a one-time event in our past. Multiple bangs went off before it in remote parts of the universe, and countless others will erupt elsewhere in the future.*

·

Once this new worldview was clear in my mind, I could not wait to share it with other cosmologists. And who better could I choose as my first confidant but Mr. Inflation himself—Alan Guth, whose office at MIT was only a 20-minute drive away from Tufts. So I did just that—I drove to the famous institute for a meeting with Alan.

MIT occupies a monstrous conglomeration of buildings, where I have gotten hopelessly lost on many occasions. You may walk on the third floor of building 6 and suddenly discover that you are already on the fourth floor of building 16. I decided to play it safe and took the simplest, although the longest, way to my destination: through the main entrance (marked by a row of Corinthian columns and crowned with a green dome). I had to march all the way along what locals call the Infinite Corridor, then climb a few flights of stairs, and finally I was in Alan Guth's office.

I told Alan about the random walk of the scalar field and how it could be described mathematically. But then, when I was in the middle of unveiling my new dazzling picture of the universe, I noticed that Alan was beginning to doze off. Years later, when I got to know Alan better, I learned that he is a very sleepy fellow. We organize a joint seminar for the Boston-area cosmologists, and at every seminar meeting Alan falls peacefully asleep a

---

*To avoid confusion, from now on I will reserve the term "big bang" for the end of inflation and use the term "singularity" for the initial (or final) state of infinite curvature and density.

few minutes after the talk begins. Miraculously, when the speaker is finished, he wakes up and asks the most penetrating questions. Alan denies any supernatural abilities, but not everybody is convinced. So, in retrospect, I should have continued the discussion. But at the time I was not aware of Alan's magical powers and hastily retreated.

The response I got from other colleagues was also less than enthusiastic. Physics is an observational science, they said, so we should refrain from making claims that cannot be observationally confirmed. We cannot observe other big bangs, nor can we observe distant inflating regions. They are all beyond our horizon, so how can we verify that they really exist? I was disheartened by such a cool reception and decided to publish this work by embedding it as a section in a paper on a different subject: I came to think that it did not deserve to have a full paper devoted to it.[1]

To explain the idea of eternal inflation in the paper, I used the analogy of a drunk wandering near the top of a hill. In a couple of months I got a letter from the editor saying that my paper was accepted, except that the discussion of drunks "was not appropriate for an archival journal like the *Physical Review*" and I should replace it with a more suitable analogy. I heard of a similar incident that happened earlier to Sidney Coleman. He had a diagram in his paper that looked like a circle with a wiggly tail. Coleman referred to it as a "tadpole diagram." Predictably, the editor complained that the term was inappropriate. "OK," replied Coleman, "let us call it sperm diagram." Following that, the original version of the paper was accepted without further comment. I briefly contemplated using Coleman's tactic, but then decided against it and removed the drunk analogy altogether: I did not feel like picking a fight.

I did not return to the theory of eternal inflation for nearly ten years. Except for one episode . . .

A GLIMPSE OF ETERNITY

I went on to work on my other research interests, and at times it even appeared that I was cured of my strange obsession with unobservable worlds. But the truth was that the temptation to get a glimpse of the universe beyond the horizon did not go away. In 1986, when I could not resist it any more, I developed a computer simulation of the eternally inflating universe, together with a graduate student, Mukunda Aryal.

I am technologically challenged and have never written a single line of computer code. But I have a pretty good understanding of how computers think, and over the years I have supervised several major computational projects by graduate students. Since I cannot check the code (and I suspect that I would not enjoy that even if I could), I am wary of hidden dangers and always view the results with great suspicion. So I made Mukunda go through multiple checks and run the simulation for simple cases, where we knew the answers. Finally, when I was satisfied that everything worked fine, we turned to the real thing.

We started the simulation with a small region of false vacuum, represented by a light square area on the computer screen. After a while, the first dark islands of true vacuum made their appearance. These island universes grew rapidly in size, as their boundaries advanced into the inflating sea. But the inflating regions expanded even faster, so the gaps separating the island universes widened and new island universes formed in the newly created space.[2]

The picture that emerged after we allowed the simulation to run for some time showed large island universes surrounded by smaller ones, which were surrounded by still smaller ones, and so forth, resembling an aerial view of an archipelago—a pattern known as a fractal to mathematicians. Figure 8.3 is the result of a similar, but more sophisticated, simulation that I later developed with my students Vitaly Vanchurin and Serge Winitzki.

Mukunda and I published the results of our simulation in the European journal *Physics Letters*.[3] My curiosity about the unobservable universe now satisfied, I moved on to other things. In the meantime, the subject was taken up by Andrei Linde.

LINDE'S CHAOTIC INFLATION

Linde is the hero of inflation, the person who saved the theory by inventing a flattened energy hill for the scalar field. Since 1983, he was developing the idea that the universe started in a state of primordial chaos. The scalar field in that state was varying wildly from one place to another. In some re-

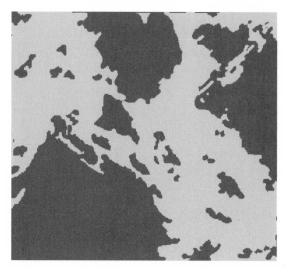

Figure 8.3. Computer simulation of an eternally inflating universe. Island universes (dark) in the inflating false-vacuum background (light). The larger island universes are the older ones: they had more time to grow.

gions it happened to be high up on the energy hill, and that's where inflation took place.

Linde realized that it was not necessary for the field to start at the highest point of the energy landscape. It could also roll down from some other point on the slope. In fact, the energy hill might have no highest point and keep rising without limit (see Figure 8.4). Such a "topless" hill has a true vacuum at the bottom, but there is no definite location for the false vacuum. Its role can be played by any point on the slope where the field happens to be in the initial chaos, provided that it is high enough to allow sufficient roll-down time for inflation. Linde described these ideas in a paper called "Chaotic Inflation."

A few years later, Linde investigated the effect of quantum kicks on the scalar field in this scenario. Surprisingly, he found that they can also result in eternal inflation, even though the energy hill does not have a flat top. His key observation was that at higher elevations quantum kicks get stronger and can push the field up, against the downward force of the slope. So, if the field starts high up the hill, it pays little attention to the slope and undergoes

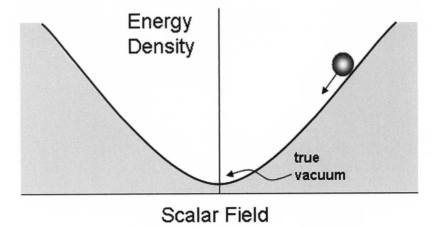

Figure 8.4. Scalar field rolls down the slope of a "topless" energy hill.

a random walk, as in the flat hilltop case. When it wanders into the lower terrain, where quantum kicks are weak, it rolls in an orderly way down to the true vacuum. The time it takes for this to happen is much longer than the inflationary doubling time; so inflating regions multiply faster than they decay, which results once again in eternal inflation.

Here I should stop to clarify the terminological confusion that bedevils this subject. "Eternal inflation" is often confused with "chaotic inflation," although the two are very different. "Chaotic" refers to a chaotic initial state and has nothing to do with the eternal character of inflation. Linde showed that chaotic inflation can also be eternal, but that's where the connection ends. For clarity, in the rest of this book I will limit the discussion to the original inflation model with a flattened energy hill. Eternal inflation on a topless hill is similar.

Linde's paper on eternal inflation was published three years after mine and was met with as much enthusiasm.[4] But his reaction was different. He stuck to his guns, continued this line of research, and gave numerous talks on the subject. Still, the physics community was not swayed by his efforts. It took nearly two decades before the fortunes of eternal inflation started to turn.

9

The Sky Has Spoken

What is now proved was once only imagined.

—William Blake

T he theory of inflation was little more than a speculative hypothesis when Alan Guth proposed it in 1980. But by the end of the 1990s it was well on its way to becoming one of the cornerstones of modern cosmology. New observational data were coming in, confirming the predictions of the theory, at times in a rather unexpected way.

RETURN OF THE COSMOLOGICAL CONSTANT

The most straightforward prediction of inflation is that the observable region of the universe should have a flat, Euclidean geometry. The universe as a whole may well be spherical, or have a more complicated shape, but our horizon encompasses only a tiny part of it, so we cannot distinguish it from flat. As we discussed in Chapter 4, this statement is equivalent to saying that the average density of the universe should be equal to the critical density with a very high accuracy.

In the early days of inflation, astronomers viewed this prediction with a high degree of skepticism. Ordinary matter, consisting of protons, neutrons, and electrons, adds up to only a few percent of the critical density. There is also a much larger amount of what is called *dark matter*, made up of some unknown particles. As its name suggests, the dark matter cannot be seen directly, but its presence is manifested by the gravitational pull it exerts on visible objects. Observations of how stars and galaxies move indicate that the mass in dark matter is about ten times greater than that in ordinary matter. Still, putting it all together, the total mass density of the universe comes out to be about 30 percent of the critical density, 70 percent short of the target.

This is where things stood until 1998, when two independent teams announced a startling discovery.[1] They measured the brightness of supernova explosions in distant galaxies and used the data to figure out the history of cosmic expansion.* To their great surprise, they found that instead of being slowed down by gravity, the speed of expansion is actually accelerating. This finding suggests that the universe is filled with some gravitationally repulsive stuff. The simplest possibility is that the true vacuum, which we now inhabit, has a nonzero mass density.† As we know, vacuum is gravitationally repulsive, and if its density is greater than half the average density of matter, the net result is repulsion.

The mass density of the true vacuum is what Einstein called the cosmological constant—the idea he denounced as his greatest blunder. It lay buried for nearly seventy years, but now it looks as though it was not such a bad idea after all. As we shall see later in this book, the sudden return of the cosmological constant led to a deep crisis in elementary particle physics. But for the theory of inflation it was a very welcome development. The mass density of the vacuum, evaluated from the rate of cosmic acceleration, amounts to about 70 percent of the critical density—precisely what is needed to make the universe flat!

This conclusion was later independently confirmed by observations of the cosmic microwave radiation. Rather than relying on Friedmann's link

---

*The distance to a supernova, which is determined from how bright it appears as viewed from Earth, tells us how long its light has traveled and, thus, when the explosion occurred. The reddening of the light (the Doppler redshift) can then be used to evaluate the speed of cosmic expansion at that time. More on this in Chapter 14.

†Some other options will be mentioned in the following chapters. Many physicists take an agnostic attitude toward the cause of cosmic acceleration and refer to it as "dark energy."

between the geometry of the universe and its density, the microwave observations probe the geometry directly—in essence, by measuring the sum of the angles in a huge narrow triangle with one vertex on Earth and the other two at the points of emission of microwaves arriving to us from two nearby directions in the sky. (The longer sides of this triangle have lengths of about 40 billion light-years.) In flat space, the angles should add up to 180 degrees, as you might remember from your geometry class at school. A greater value of the sum of three angles would indicate a closed universe of spherical geometry (see Figure 9.1), and a smaller value would point to an open universe with the geometry of a saddle. The microwave observations showed that the sum of the angles is in fact very close to the flat-space answer. These results can be re-expressed in terms of the densities, using Friedmann's geometry-density relation. The most recent measurements then imply that the density of the universe is equal to the critical density with an accuracy of better than 2 percent—a spectacular success for inflationary cosmology.

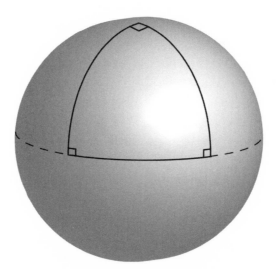

Figure 9.1. In a spherical universe, the sum of the angles in a triangle is greater than 180 degrees. The triangle in this figure has three right angles, which add up to 270 degrees.

IMAGES OF THE BLAZING PAST

Another triumph of inflation has been the explanation of small-density perturbations, the tiny ripples that later evolved into galaxies. The theory of inflation makes a sharp prediction—that the magnitude of perturbations should be nearly the same on all astrophysical distance scales, from the typical interstellar distance (a few light-years) all the way to the entire visible universe. By the early 1990s the observers were ready to put this prediction to a test.

As we discussed in Chapter 4, the primordial ripples leave an imprint on the cosmic background radiation. This afterglow of the big bang was emitted more than 13 billion years ago and now comes to us from all directions in the sky. Ever since its discovery in the mid-1960s, cosmologists were aware that hidden in this radiation was an image of the early universe. However, the primordial non-uniformities are so small, only one part in 100,000, that for many years they were beyond the accuracy of the measurements, and all one could observe was a perfectly uniform background. The breakthrough occurred in 1992, with the launch of the Cosmic Background Explorer (COBE) satellite. COBE produced a full map of the sky, detecting radiation from every direction, and we were, for the first time, able to discern tiny variations in the intensity of the radiation.

The COBE map is like a photograph that is somewhat out of focus: it captures the gross features of the cosmic fireball, but finer details, smaller than about 7 degrees on the sky, are completely blurred. (For comparison, the Moon subtends an angle of about half a degree.) COBE was followed by a series of other experiments, of ever increasing accuracy. The most recent of these was another satellite mission, WMAP.* Its image of the fireball, shown in Figure 4.2, resolves features as small as one-fifth of a degree; it is thirty times sharper than COBE's original map.

Step by step, as the data accumulated, the pattern of primordial ripples gradually emerged. And amazingly, it was in striking agreement with the predictions of inflation! These records of the hot early epoch were there in

*See the second footnote on p. 42.

the sky for billions of years, waiting to be discovered and deciphered. Now, finally, the sky has spoken.

In the years to come, the theory of inflation will face a succession of new observational tests. A physical theory can be supported by the data, but it can never be proved. On the other hand, a single well-established fact that contradicts the theory would be enough to disprove it. For example, inflation predicts that the density should be equal to the critical density with an accuracy of 1 in 100,000. So, if some future experiment discovers a greater deviation from the critical density, inflation will be in trouble.[2]

The next-generation microwave background missions include the Planck satellite,* which will further improve the image resolution, and the ground-based Clover and QUIET observatories. Clover and QUIET will accurately measure the orientation of the electric field, or *polarization*, of the microwaves. The polarization pattern is sensitive to the presence of gravitational waves—tiny vibrations of spacetime geometry. This effect can be used to test yet another prediction of inflation: we should be bathing in gravitational waves with a very wide spectrum of lengths, ranging from less than the size of the solar system up to the largest observable scales.[3] The amplitude of the waves is determined by the energy of the false vacuum that drives inflation: the higher the vacuum energy, the larger the waves. Thus, if Clover detects gravitational waves, we should be able to deduce the energy of the false vacuum that drove the inflationary expansion.[4] This would be an important step in our understanding of inflation and of its connection with the physics of the microworld.

•

As the new data were coming in, my thoughts were going back to my neglected brainchild, the theory of eternal inflation. The main objection against it was that it was concerned with the universe beyond our horizon, which is not accessible to observation. But if the theory of inflation is supported by the data in the observable part of the universe, shouldn't we also believe its conclusions about the parts that we cannot observe?

---

*The Planck satellite is named after one of the discoverers of quantum mechanics, Max Planck, who also derived a formula describing how the energy of thermal radiation is distributed between waves of different frequency. The satellite is scheduled to be launched in 2007.

If I drop a stone into a black hole, I can use general relativity to describe how it falls toward the center and how it is crushed and vaporized by immense gravitational forces. None of this can be observed from the outside, because neither light nor any other signal can escape from the black hole interior. And yet very few physicists would question the accuracy of my description. We have every reason to believe that general relativity applies inside black holes just as much as it does outside. The same case could now be made for the theory of inflation. We should try to extract from this theory as much as it will tell us about the grand design of the universe, its origin, and its ultimate fate.

· 10 ·

Infinite Islands

I could be bounded in a nutshell, and count myself a king of
infinite space . . .   —SHAKESPEARE, *Hamlet*

THE FUTURE OF CIVILIZATIONS

The question that started me thinking about eternal inflation again had more to do with science fiction than with physics. It was about the future of intelligent life in the universe. The long-term prospects for any civilization appear to be rather bleak. Even if a civilization avoids natural catastrophes and self-destruction, it will, in the end, run out of energy. The stars will eventually die, and all other sources of energy will also come to an end. But now eternal inflation appeared to offer some hope.

Stars will die in our cosmic neighborhood, but an infinite number of new stars will form in the future big bangs of eternal inflation. Our visible region is but a tiny part of one island universe, lost in the inflating sea of false vacuum (see Figure 8.3). New island universes constantly emerge in the midst of that sea, bringing in myriads of new stars. In fact, star formation will always continue even within our own island universe.

The frontiers of island universes are constantly advancing into the inflating sea. This relentless advance is caused by the decay of false vacuum in the adjacent inflating regions. These frontiers are thus the regions where the big bang is happening right now.* Newly formed island universes are microscopically small, but they grow without limit as they get older. Central parts of large island universes are very old. They are dark and barren: all stars have long since died there, and life has become extinct. But regions at the periphery of the islands are very new and must be teeming with shining stars.

Advanced civilizations may wish to send missions to colonize newly formed stellar systems near the boundary of their island. If not, they could at least send messages to new civilizations that will evolve close to the boundary, or in other island universes. Those civilizations could in turn send messages to posterity, and so on. If we follow this path, we could become a branch in an ever-growing "tree" of civilizations and our accumulated wisdom would not be completely lost.

These scenarios were suggested by Andrei Linde in a paper called "Life after Inflation,"[1] and I wanted to know if any of them is actually possible, at least in principle. Linde analyzed various aspects of the problem, but did not commit himself to a definite answer. The fact that stars in some part of the universe are formed later than they are formed here does not necessarily mean that we can get from here to there in the available time. Besides, we know from Einstein that the notions of "earlier" and "later" are not absolute and may be observer-dependent. To make any progress with the problem, I had to understand the spacetime structure of the eternally inflating universe.

As we discussed in Chapter 2, space and time in the theory of relativity are united in a four-dimensional entity called spacetime. A point in spacetime is an *event*, which has a certain location and time. Consider, for example, two events that you may wish to attend. One is your class reunion here on Earth and the other is an interstellar superball game, which is scheduled to take place three years later at the star Alpha Centauri, about four light-years away from here. The question is, Can you get to both of these events?

The answer can be found by calculating the *spacetime interval* between the two events. The interval between events in spacetime plays the role analogous to the distance between points in space. Its mathematical definition is not important for us here; what is important is that the interval can be of

---

*Remember that we agreed to identify the big bang with the end of inflation.

two kinds: it is either *spacelike* or *timelike*. The interval is timelike if a material object can get from one event to the other without violating the basic tenet of relativity—that it should not move faster than the speed of light.[2] In this case all observers will agree on which of the two events is earlier and which is later. Alternatively, if getting from one event to the other is impossible (that is, if it requires faster-than-light motion), the interval is spacelike. None of these events can then be caused by the other. Einstein showed that the time order of such events is observer-dependent and that there always exists an observer who will find that they occurred simultaneously.

In our example with Alpha Centauri, the interval turns out to be spacelike, so you will have to choose which of the two events you want to attend. In fact, in this example it is easy to figure out the answer without calculating the interval. The distance traveled by light in three years is 3 light-years; so in order to cover the 4-light-year distance to Alpha Centauri, you would have to move faster than light. In the curved spacetime of the eternally inflating universe, the analysis is more complicated, and one does have to calculate the interval.

The spacetime of an island universe is schematically illustrated in Figure 10.1. The vertical direction is time, and the horizontal direction is one of the three spatial dimensions; the other two dimensions are not shown. Each horizontal line gives a snapshot of the universe at a moment of time. You can follow the history of the island universe by starting with a horizontal dotted line marked "before" at the bottom of the figure and gradually moving it upward. (The moment of time represented by this line is in the inflating part of spacetime, where the island universe has not yet formed.) The thick solid line labeled "Big Bang" is the boundary between the island universe and the inflationary part of spacetime. The location marked by a black galaxy is the here and now, and white galaxies mark spacetime regions where the conditions are similar to what we have here today. The horizontal dotted line labeled "now" represents the present time. It shows the island universe with a barren central region and some star-forming regions close to the boundaries.

A simple calculation showed that all big bang events, which are located along the solid line in the figure, are separated by spacelike intervals. That was the key observation; it gave me the answer to my question about the fu-

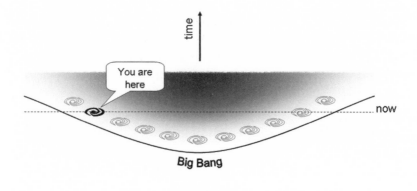

Figure 10.1. Spacetime diagram of an island universe (global view).

ture of civilizations. It also completely changed the way I viewed the island universes.

The spacelike character of the intervals means that you cannot get from any one of the big bang events to any other. In other words, you cannot keep up with the expanding boundaries of the island universe: they are expanding faster than the speed of light. Thus, we will never be able to reach the shores of the inflating sea and bask in the light of the new suns that will be born there. We cannot even send any messages to the future civilizations that will thrive around these suns, since no signal can travel faster than light. Regrettably, eternal inflation does not seem to improve the long-term prospects for humanity.

You may be puzzled by faster-than-light expansion of island universes, as it apparently contradicts Einstein's ban on superluminal velocities. The ban, however, is very specific: it applies only to the motion of material objects (including radiation, such as light or gravitational waves) relative to one another, while the boundary of an island universe is a geometric entity, which does not have any mass or energy.

The faster-than-light expansion of the boundary means that successive big bangs cannot be causally related. They are not like a row of dominoes, where the fall of one domino triggers the fall of the next. The progression of false-vacuum decay is predetermined by the pattern of the scalar field that was produced during inflation. The field variation in space is very gradual, and as a result the false-vacuum decay in nearby regions occurs almost simultaneously. That is why the big bangs happen in such a quick succession and the boundary is advancing so fast.

TIME IS OF THE ESSENCE

I confess to You, Lord, that I still don't know what time is.
—Saint Augustine

What do we actually mean when we say that the big bang at the boundary of an island universe occurred later than it did in the central region? Since all the big bang events are spacelike-separated, different observers will disagree on which of these events occurred earlier and which later. Whom, then, should we listen to? Whose clock should we use to time the big bangs? We shall now stop to reflect upon this issue. The analysis is somewhat intricate, but it's worth the effort, as it leads to some far-reaching implications.

As a warm-up exercise, let us first consider a homogeneous universe described by one of the Friedmann models. Homogeneity means that matter is uniformly distributed in space at any moment of time. This sounds simple, but we need to define what is meant by a "moment of time."

When cosmologists talk about a "moment of time," they picture a large number of observers, equipped with clocks, and scattered throughout the universe. Each observer can see a small region in her immediate vicinity, but the whole assembly of observers is needed to describe the entire universe. We can think of ourselves as one member in this assembly. Our clock now shows the time 14 billion years A.B.* "The same time" in another part of the universe is when the clock of the observer located there shows the same reading. We have to decide, though, how observers, who are outside each other's horizon, are to synchronize their clocks.

In the case of Friedmann's universe, the answer is simple: the big bang is the natural origin of time in that universe, so each observer should count

---

*As before, "A.B." stands for "after the big bang."

time starting from the big bang.* With this definition of simultaneity, the matter density measured by all observers at the same time will be the same, so the universe is homogeneous.

We could, in principle, imagine an assembly of observers whose clocks are set up differently. For example, we could offset the origin of time by some amount away from the big bang and make this amount vary from one region of space to another. The universe would then look very complex and inhomogeneous. Of course, no one in his right mind would use such a description. It merely complicates matters and conceals the true nature of Friedmann's universe. But things are not always so straightforward.

Going back to the eternally inflating universe, let us first consider a large region, like the one shown in Figure 8.3, which includes both island universes and inflating domains. There is no obvious choice for the origin of time in such a region. The definition of a "moment of time" is therefore largely arbitrary, the only condition being that all events at that "moment" should be separated by spacelike intervals. Once the choice is made for one such moment, the clocks of the observers are set and the notion of time is defined for the whole subsequent history of the region. If we choose the initial moment early enough, when the entire region is in the false-vacuum state, then at later times island universes will appear and expand, as we discussed in the preceding section. But the order of their appearance and the pace and pattern of their expansion can be rather different for different choices of the initial "moment."

Suppose now that we are interested in one specific island universe and want to describe it from the point of view of its inhabitants. The situation is then entirely different. As in the case of the Friedmann universe, there is now a natural choice for the origin of time. All observers inhabiting the island universe can count time from the big bang at their respective locations. In other words, the big bang can be chosen as the initial "moment of time." This choice leads to a new, and drastically different, picture of the island universe. To distinguish between the large-region and single-island descriptions, we shall refer to them as "global" and "local" (or "internal") views, respectively.

---

* The state of motion of the observers also affects the readings of their clocks. In a Friedmann universe, it is most natural to assume that the observers are at rest relative to galaxies (or matter particles) at their respective locations. These are the "co-moving" observers.

The internal view of the island universe is illustrated in the spacetime diagram of Figure 10.2. As in Figure 10.1, the moment of the big bang is represented by a solid curve marked "Big Bang." The density of matter at all the big bang events on this curve is very nearly the same, as it is determined by the density of the decaying false vacuum. Thus, in the local view, the island universe is nearly homogeneous. The present moment in this view is represented by the dotted line marked "now," which coincides with the line of galaxies in the figure. All points on this line are characterized by the same average density of matter and the same density of stars as observed in our local part of the universe. But most remarkably, from the internal point of view the island universe is infinite!

In the global view, the island universe grows with time, as new big bangs go off at its boundary, and gets arbitrarily large if you wait long enough. But in the local view, the big bangs happen all at once and the island universe is infinitely large from the very beginning. In Figure 10.2, this infinity is reflected in the fact that the solid line representing the big bang never comes to an end. Extensions of this curve correspond to later and later big bangs in the global view and to more and more distant regions at the initial moment in the local view. The infinity of time in one view is thus transformed into the infinity of space in the other.

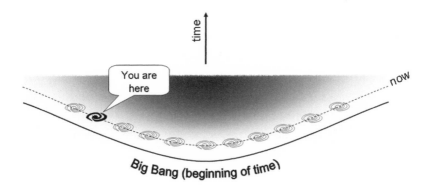

Figure 10.2. Internal view of the island universe spacetime.

THE BIG PICTURE

Let us now briefly summarize what we have learned about eternal inflation. If we were somehow able to observe the eternally inflating universe from outside, as the surface of the Earth can be observed from outer space, we would see a multitude of island universes scattered in the vast inflating sea of false vacuum. If the universe is closed, then the view that would open in front of us might in fact resemble a picture of the globe, with continents and archipelagos surrounded by the ocean.* This globe is expanding at a staggering speed, the island universes are also growing exceedingly fast, and tiny new islands constantly appear and immediately start to expand. The number of island universes rapidly grows with time. It grows without bound and becomes infinite in the limit of infinite future.

The inhabitants of island universes, like us, see a very different picture. They do not perceive their universe as a finite island. For them it appears as a self-contained, infinite universe. The boundary between this universe and the inflating part of spacetime is the big bang, a moment in their past. We cannot travel to the inflating sea, simply because it is impossible to travel to the past.

Remarkably, the entire "master" universe, which contains all the infinite island universes, may be closed and finite. The apparent contradiction is resolved by the fact that the internal notion of time in island universes is different from the "global" time that one has to use to describe the entire spacetime. In the global time, the outer parts of island universes are not yet formed and will complete their formation in the infinite future, while in the interior time the island universe forms all at once. The spacetime structure of a closed, eternally inflating universe is illustrated in Figure 10.3.

The surprising feature of island universes—that they are infinite when viewed from the inside—turned out to be important. It later led me to what perhaps is the most striking consequence of eternal inflation.

---

*Except, of course, that a closed universe is like a three-dimensional sphere, while the surface of the Earth has two dimensions.

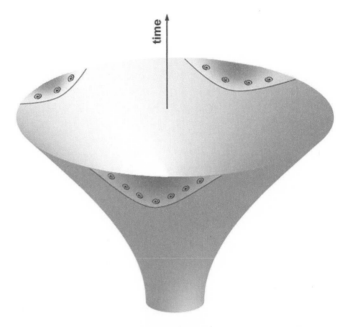

Figure 10.3. Spacetime of a one-dimensional, closed, eternally inflating universe. This universe is filled with false vacuum at the initial moment (bottom of the figure) and has three island universes by the time corresponding to the top of the figure.

11

The King Lives!

Must not all things that can happen have already happened,
been done, run past? —NIETZSCHE

CADAQUÉS

The first glimpse of an idea came to me in the summer of 2000. As it often happens, my immediate impulse was to share it with someone. You may get more credit if you work alone, but working together is so much more fun! And if you are blessed to have a good collaborator, it can be a real joy. By a stroke of luck, my old friend Jaume Garriga happened to be in town. When I told him about my thought, he understood it instantly.

Jaume is a soft-spoken, quiet fellow. He does not say much, but he does speak his mind. On that occasion, he only said, "This is a very marketable idea." That was not exactly an endorsement. He meant that it was the kind of idea that would be more attractive to mass media than to physicists. But I could tell that Jaume was hooked. He was about to leave for his native Catalonia, and we agreed to resume the discussion during my visit to the University of Barcelona, where he worked.

Two months later, Jaume met me and my wife at the Barcelona airport. We arrived on a weekend and had two days free before the beginning of my "official" visit. I could not wait to get back to our physics discussion, but it turned out that our program had already been fixed. As we drove out onto

Figure 11.1. Jaume Garriga. (Courtesy of Takahiro Tanaka)

the highway, Jaume told us that we were going to his father's farm: "They are expecting us for dinner." We passed the formidable mountain mass of Montserrat, which suddenly rises out of flat, reddish terrain, and continued north, into a greener, hilly countryside. In an hour or so we drove up to the Garriga family farm.

Amazingly, the same family has worked this land for more than 750 years. The farmhouse was an impressive Catalan *masia*, looking like a small fortress, complete with a tower. I was totally blown away and forgot all about physics.

The dinner was served in a spacious hall, where the Garriga family had gathered. As a guest of honor, I was seated next to Jaume's father, who enchanted us with tales of the ancient history of the land and made sure that my wineglass was never empty. Toward the end of the dinner he excused himself and walked out of the hall. Jaume explained: "He went to tell the cows to go home." The cows did not have to be shepherded; they just needed a friendly reminder.

Figure 11.2. Jaume in his younger years at the family farm. (Courtesy of Jaume Garriga)

After the dinner, Jaume's older brother led us up the winding stairway to the top of the tower. It was used as a watchtower in times of danger. When the enemy was sighted, the guardsmen could signal with a torch to similar towers at the neighboring farms, all the way to the duke's garrison at the castle of Cardona, about five miles away. We looked out of the small square windows of the tower—to check if there were any villains in sight. The sun was already setting over the hills. In the distance, we saw the cows coming home, by themselves, from the pasture.

We left the farm in the morning and headed north, toward the mountains. Our destination was the little coastal village of Cadaqués, the home of Salvador Dalí. My wife is fascinated by Dalí's art, and she wanted to see his house and the village where he spent most of his life. She wanted to go there every time we visited Barcelona, but once I got to the university, I was invariably distracted by physics discussions and other equally important matters, so in the end there was never any time left for the trip. Now she said this was it: we were going to Cadaqués *before* Barcelona.

The narrow winding road climbed up the mountains, clinging dangerously to the slopes, and then weaved its way down to the cliffs and secluded blue coves of Costa Brava. We entered the village in the early afternoon, when the sun was in its full Mediterranean splendor. The whitewashed houses of Cadaqués crowded the hillsides, tumbling down and stopping right at the water's edge. Higher up the slope stood a rustic white church, austere and beautiful.

Our visit to Dalí's house did not go as planned. Jaume's wife, Julie, who decided to join us at the last minute, took along their baby daughter, Clara. As we were entering the museum, Clara protested loudly, so the ladies went in, while Jaume and I were left to babysit. Soon we were deep in the discussion of our physics problem. By the time our wives returned, the museum was already closing. So I did not get to see the so-much-talked-about Casa Dalí.

We spent the rest of the afternoon wandering about the village. As we strolled the narrow cobblestone streets of Cadaqués, Jaume and I continued our discussion, and a new picture of the universe was gradually taking shape. It was bizarre and disturbing.

Figure 11.3. *Port Alguer (Cadaqués)* by Salvador Dalí. (© 2005 Salvador Dalí, Gala–Salvador Dalí Foundation/Artists Rights Society [ARS], New York)

LIMITED OPTIONS

The conversation revolved around the remote regions of the universe and how they could differ from our local *cosmic* neighborhood. Since each island universe is infinite from the viewpoint of its inhabitants, it can be divided into an infinite number of regions having the same size as our observable region. For short, we called them "O-regions."

Imagine an infinite space packed with gigantic spheres, 80 billion light-years in diameter. Each sphere is an O-region. The spheres expand with the expansion of the universe; consequently, they were smaller at earlier times. All of these O-regions looked pretty much the same at the big bang—that is, at the end of inflation. But they were different in detail. Small density perturbations, brought about by random quantum processes during inflation, differed from one region to another. As these perturbations were amplified by gravity, the macroscopic properties of O-regions began to diverge. By the time of galaxy formation, the details of how galaxies are distributed in different O-regions varied considerably, although statistically the regions were still very similar. Later on, the evolution of life and intelligence was influenced by chance, leading to further divergence of properties. We can thus expect the histories of different O-regions to be rather different.

The key observation was that the number of distinct configurations of matter that can possibly be realized in any O-region—or, for that matter, in any finite system—is finite. One might think that arbitrarily small changes could be made in the system, thus creating an infinite number of possibilities. But this is not the case.

If I move my chair by 1 centimeter, I change the state of our O-region. I could instead move it by 0.9, 0.99, 0.999, etc., centimeter—an infinite sequence of possible displacements, which more and more closely approach the limit of 1 centimeter. The problem, however, is that displacements too close to one another cannot even in principle be distinguished, because of the quantum-mechanical uncertainty.

In classical, Newtonian physics, the state of a physical system can be described by specifying the positions and velocities of all its constituent particles. We now know that such a description can only be used for macroscopic, massive objects, and even then only approximately. In the quantum world, particles are inherently fuzzy and cannot be precisely localized.

At the core of quantum physics is the *uncertainty principle*, discovered by Werner Heisenberg in 1927. It says that the position and the velocity of a particle cannot both be accurately measured. The more precisely we measure the position, the greater is the uncertainty in the velocity. If the position is measured exactly, then the velocity is completely undetermined, and vice versa—if we measure velocity exactly, we have no idea where the particle is.

Heisenberg offered the following intuitive explanation for the uncertainty. A simple way to determine the location of a particle is to shine light on it. The light waves will be scattered by the particle in all directions. Some of them will be registered by our eyes, or by our measuring apparatus, and we will see where the particle is. The image of the particle obtained in this way cannot be absolutely sharp: the details smaller than the wavelength of light are necessarily blurred, so the position cannot be measured more accurately than the light's wavelength. To deal with this problem, we could use light of shorter and shorter wavelengths, but that's where the quantum nature of light comes into play. Light consists of photons whose energy is inversely proportional to the wavelength. When the particle is illuminated with very-short-wavelength light, it is being bombarded by highly energetic photons. The particle recoils from the impact, and thus its velocity is altered. This recoil is the origin of the uncertainty: the greater the accuracy we want to achieve in position measurement, the shorter-wavelength light we should use, and the greater will be the impact on the velocity of the particle we are observing.

Even if we are not interested in the particle's velocity and want to know only its position, Heisenberg's argument indicates that in order to localize the particle more and more accurately, we would have to supply larger and larger amounts of energy. In any realistic physical system with limited energy, the localization accuracy is also limited.

Since we cannot pin down precise positions of particles, we can instead use what is called the coarse-grained description. Suppose the volume of our O-region is divided into cubic cells of a certain size, say, 1 cubic centimeter each. A coarse-grained state is given by indicating the cell occupied by each particle in the region. A more refined description is obtained by making the cells smaller. But there is a limit to this refinement, since the energy cost of localizing particles to small cells will eventually exceed the available energy in the O-region.

Clearly, the number of ways in which a finite number of particles can be distributed into a finite number of cells is also finite. Hence, the material content of our O-region can only be in a finite number of distinct states. A very rough estimate of this number gives 10 to the power 10^{90}, or 1 followed by 10^{90} zeros—far too many zeros to fit in the pages of this book. This is a fantastically huge number, but the important point is that the number is finite.

So far, so good. One problem though is that some distant regions may contain more matter and energy than ours. Rare, large quantum fluctuations during inflation may produce some strongly over-dense regions, full of high-energy particles. As the number of particles and their energy grow, the number of possible states is also increased. But only up to a point. If more and more energy is packed into a region, its gravity gets stronger, and eventually the whole region turns into a black hole. Thus gravity puts an absolute bound on the number of states that can possibly exist in a region of a given size, regardless of its contents.

The precise value of the bound is still a subject of investigation. It was originated by Jacob Bekenstein in the 1980s and was taken up more recently, in the context of superstring theory, by Gerard 't Hooft, Leonard Susskind, and others. This work suggests that the maximum number of states in a region depends only on the surface area of its boundary. For an O-region, this number is 10 to the power 10^{123} (1 followed by more than a google zeros!).*

COUNTING HISTORIES

Not only is the number of possible states of an O-region finite, the number of its possible histories is finite as well.

A *history* is described by a sequence of states at successive moments of

---

*This bound does not apply to regions much greater than the cosmic horizon. It is expected to be marginally valid for an O-region, which has the same size as the horizon.

time. The notions of which histories are possible are very different in quantum and classical physics. In the quantum world, the future is not uniquely determined by the past. The same initial state can lead to a multitude of different outcomes, and we can only calculate their probabilities. As a result, the range of possibility is greatly enlarged. But once again, the quantum uncertainty does not allow us to distinguish between histories that are too close to one another.

A quantum particle does not generally have a well-defined history. This is not surprising, since, as we know, it does not have a definite position. But the uncertainty does not amount to simply not knowing which path the particle followed between its emission and its subsequent detection. The situation is much weirder: the particle appears to follow many different paths at once, and all of them contribute to the final outcome.

This schizophrenic behavior is best illustrated in the famous double-slit experiment (Figure 11.4). The setup consists of a light source and a photographic plate, which is blocked by an opaque screen with two narrow parallel slits in it. The light that gets through the slits creates an image on the plate. The experiment was first performed in the early 1800s by the English physicist Thomas Young. He found that the image displays a pattern of al-

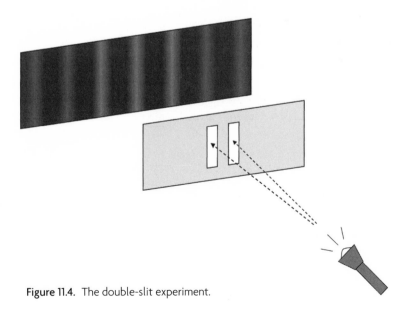

Figure 11.4. The double-slit experiment.

ternating bright and dark fringes. All points on the photographic plate receive light from both slits. But in some places the light waves arrive "in
phase" (the crests and troughs of the two waves coincide) and reinforce one
another, while in other places they are out of phase (the crest of one wave
coincides with the trough of the other) and mutually cancel. Thus the pattern of fringes is explained by the wavelike nature of light.

The weird part begins when we reduce the intensity of the source to
the point that it emits individual photons one by one. Each photon makes a
small spot on the photographic plate. Initially the spots seem to appear at
random, but remarkably, after a while, a pattern begins to build up that exactly coincides with the pattern of fringes we had before. The photons arrive at the screen separately, so photons that passed through one of the slits
have no way of interacting with photons that passed through the other.
How, then, do they manage to "reinforce" or "cancel" one another?

To probe this strange behavior a bit further, we can try to see what happens if we force photons to go through one slit or the other. Suppose we run
the experiment first with only one slit open and then with the other open
for equal amounts of time and without changing the photographic plate in
between. Since photons pass through the apparatus individually, this should
make no difference, and we should get the same pattern. Right? Wrong. In
this modified version of the experiment, no fringes are observed and the
photograph shows only the outlines of the two slits.

It follows that the picture of a photon that goes through one of the slits
and does not care whether or not the other one is open cannot be right.
When both slits are open, the photon somehow "feels" the two possible histories that it may follow. They jointly determine the probability for the photon to hit a particular spot on the plate. This phenomenon is called *quantum
interference* between the histories.

Quantum interference is rarely as apparent as in the double-slit experiment, but it affects the behavior of every particle in the universe. As they move
from one place to another, particles "sniff" many different routes, so instead
of a well-defined past, we have a tangled web of interfering histories.

How, then, can we be sure that some event has actually happened? How
can we make any sense of the concept of history? The answer, once again,
lies in the coarse-grained description.

As before, we divide space into little cells and define coarse-grained states of the system (O-region in our case) by specifying the cell "addresses" for all particles. A coarse-grained history is given by a sequence of such states at regular time intervals, say, every 2 seconds. Now, the key point is that the effect of interference is usually strong only for histories that are very close to one another. If we increase the cell sizes and time intervals, different coarse-grained histories become more and more distinct from one another, and at some point their interference becomes completely negligible. We can then meaningfully talk about alternative histories of the system.

The formulation of quantum mechanics in terms of alternative coarse-grained histories was developed relatively recently, in the 1990s, by Robert Griffiths, Roland Omnes, James Hartle, and Murray Gell-Mann. They found, in particular, that the minimum size of cells that is still consistent with a definite history is typically microscopic and that the minimum time interval is a tiny fraction of a second. Not surprisingly, history is a well-defined concept in the macroscopic world of human experience.

A coarse-grained history proceeds in finite time steps, and a history of any finite duration must consist of a finite number of moments. At each moment, the system can only be in a finite number of states, and it follows that the number of distinct histories of the system must also be finite.

Jaume and I did a quick, back-of-the-envelope estimate of the number of possible histories that can occur in an O-region from the big bang till the present. As one might expect, we got yet another "googleplexic"* number: 10 to the power 10^{150}. The actual numbers of quantum states and of histories in an O-region are not particularly important, but the finiteness of these numbers has profound implications, as we shall now discuss.

HISTORY REPEATS ITSELF

Let us now take stock of the situation. It follows from the theory of inflation that island universes are internally infinite, so each of them contains an infinity of O-regions. And it follows from quantum mechanics that there is only a finite number of histories that can unfold in any O-region. Putting these two statements together, we arrive at the inevitable conclusion that every single history should be repeated an infinite number of times. Ac-

---

*From "googleplex"—the name for the number 10 to the power 10^{100}.

cording to quantum mechanics, anything that is not strictly forbidden by conservation laws has a nonzero probability of happening. And any history that has a nonzero probability will happen—or rather has happened—in an infinite number of O-regions!

Included among these infinitely replayed scripts are some very bizarre histories. For example, a planet similar to our Earth can suddenly collapse to form a black hole. Or it can emit a huge pulse of radiation and switch to another orbit, much closer to the central star. Such occurrences are extremely unlikely, but this only means that one will have to survey a very large number of O-regions before encountering one of them.

A striking consequence of the new picture of the world is that there should be an infinity of regions with histories absolutely identical to ours. Yes, dear reader, scores of your duplicates are now holding copies of this book. They live on planets exactly like our Earth, with all its mountains, cities, trees, and butterflies. The earths revolve around perfect copies of our Sun, and each sun belongs to a grand spiral galaxy—an exact replica of our Milky Way.

How far away are these earths populated by our duplicates? We know that matter contained in our O-region can be in about 10 to the 10^{90} different states. A box containing, say, a googleplex (10 to the 10^{100}) O-regions should exhaust all these possibilities, with a large margin. Such a box should be, roughly, a googleplex light-years across. At larger distances, O-regions, including ours, should recur.

There should also be regions where histories are somewhat different from ours, with all possible variations. When Julius Caesar stood with his legions on the bank of the river Rubicon, he knew he was about to make a momentous decision. Crossing the river would amount to high treason, and there would be no way back. With the words *"Iacta alea est!"*—"The die is cast!"—he ordered the troops to advance. And the die was cast indeed: on some earths Caesar went on to become the dictator of Rome, while on others he was defeated, tried, and executed as an enemy of the state. Of course, on most earths there has never been a person by the name Caesar, and most places in the universe are nothing like our Earth—since there are many more ways for things to be different than for them to be the same.

It may be fitting that this surreal picture of the world originated in the town haunted by the spirit of Salvador Dalí. Like Dalí's paintings, it blends weird, nightmarish features with recognizable reality. It is, however, a direct consequence of the inflationary cosmology. Jaume and I wrote a paper describing the new worldview and submitted it to the *Physical Review*, the leading physics journal. We ran the risk that the paper could be rejected for being "too philosophical," but it was accepted without a glitch. In the discussion section at the end of the paper we wrote:

> The existence of O-regions with all possible histories, some of them identical or nearly identical to ours, has some potentially troubling implications. Whenever a thought crosses your mind that some terrible calamity might have happened, you can be assured that it *has* happened in some of the O-regions. If you narrowly escaped an accident, then you were not so lucky in some of the regions with the same prior history. . . . On the positive side, . . . some readers will be pleased to know that there are infinitely many O-regions where Al Gore is President* and—yes!—Elvis is still alive.[1]

The press responded instantly—as Jaume had anticipated. The next month's issue of the British magazine *New Scientist* published a review of our paper under the headline "The King Lives!"

WHAT ELSE IS NEW?

We later learned that the picture of multiple clones of ourselves scattered throughout the universe had some lineage. The famous Russian physicist Andrei Sakharov expressed a similar idea in his 1975 Nobel Peace Prize lecture. He said, "In infinite space many civilizations are bound to exist, among them societies that may be wiser and more 'successful' than ours. I support the cosmological hypothesis which states that the development of the universe is repeated in its basic characteristics an infinite number of times."[2]

Some people even argued that it was self-evident that absolutely every-

---

*We wrote our paper in 2001, right after the contentious presidential election in the United States, when George Bush won over Al Gore by a very narrow margin.

thing must happen in an infinite universe. This claim, however, is false. Consider, for example, the sequence of odd numbers 1,3,5,7, . . . The sequence is infinite, but you cannot conclude that it contains all possible numbers. In fact, all even numbers are missing from the sequence. Similarly, infinity of space does not, by itself, guarantee that all possibilities are realized somewhere in the universe. We could, for example, have the same galaxy endlessly repeated in the infinite space.

This point was recognized by the South African physicists George Ellis and G. Brundrit.[3] They assumed that the universe is infinite and argued that it should contain an infinity of places very similar to our Earth. (Their analysis was based on classical physics, so they could only argue that other earths were similar, but not identical, to ours.) They had to assume in addition that the initial state of the universe varied randomly from one O-region to another, so that all possible initial states were exhausted in the infinite volume. Thus, the existence of our clones is not certain, but hinges on the assumptions of spatial infinity and the "exhaustive randomness" of the universe.

In contrast, in eternal inflation these features do not have to be introduced as independent assumptions. It follows from the theory that island universes are infinite and that the initial conditions at the big bang are set by random quantum processes during inflation. The existence of clones is therefore an inevitable consequence of the theory.

THE MEANING OF THE WORD "IS"

It depends on what the meaning of the word "is" is.

—BILL CLINTON

The idea of many worlds or "parallel" universes has also been discussed in a totally different context. You might have heard of the many-worlds interpretation of quantum mechanics, which asserts that the universe is constantly splitting into multiple copies of itself, with all possible outcomes of every quantum process being realized in different copies. This may sound similar to eternal inflation, but the two theories are in fact completely different. To make sure they do not get confused, let us now make a brief detour into the world of many worlds.

Quantum mechanics is a phenomenally successful theory. It explains the structure of atoms, the electric and thermal properties of solids, nuclear reactions, and superconductivity. Physicists rely on it with complete confidence—and yet, the foundations of this theory are notoriously obscure, and debate about its interpretation is still ongoing.

The most contentious issue is the nature of quantum-mechanical probabilities. The *Copenhagen interpretation*, developed by Niels Bohr and his followers, holds that the quantum world is inherently unpredictable. According to Bohr, it is meaningless to ask where a quantum particle is, unless you perform a measurement to find this out. The probabilities for all possible outcomes of the measurement can be calculated using the rules of quantum mechanics. It appears that the particle "makes up its mind" and jumps to a certain position at the last moment, when the measurement is performed.

An alternative interpretation was proposed by Hugh Everett III in his Princeton doctoral thesis in the 1950s. He argued that each possible outcome of every quantum process is actually realized, but they all occur in different, "parallel" universes. With every measurement of a particle's position, the universe branches into myriads of copies of itself, where the particle is found to be in all possible places. The branching process is fully deterministic, but we don't know which of the branches is going to be the branch of *our* experience. Thus, the outcome of *our* measurement is still subject to the law of probability, and Everett showed that all the probabilities come out exactly the same as one finds using the Copenhagen interpretation.[4]

Since the choice of interpretation does not affect any results or predictions of the theory, most practicing physicists take an agnostic attitude toward the foundations of quantum mechanics and spend little time worrying about such issues. In the words of the particle physicist Isidor Rabi, "Quantum mechanics is just an algorithm. Use it. It works, don't worry."[5] This "shut up and calculate"[6] attitude works fine, except in quantum cosmology, where quantum mechanics is applied to the entire universe. The "orthodox" Copenhagen interpretation, which requires an external observer to perform measurements on the system, cannot even be formulated in this case: there are no observers external to the universe. Cosmologists, therefore, tend to favor the many-worlds picture.

Everett and some of his followers insist that parallel worlds are all equally real, while others believe that they are just *possible* worlds and only one universe is real.* The dispute may be purely semantic: When one says there is another, parallel universe, independent of ours, what exactly does this statement mean? As President Clinton said on a different subject, "It depends on what the meaning of the word 'is' is."[7] Parallel universes are like parallel lines: they do not have any points in common. Each of them evolves in its own, separate space and time, which cannot be penetrated from anywhere in our universe. How, then, can we tell whether they are real or merely possible?†

I should emphasize that none of this affects the worldview of eternal inflation that I described earlier in this chapter. If the many-worlds interpretation is adopted, then there is an ensemble of "parallel," eternally inflating universes, each having an infinite number of O-regions. The new worldview applies to each of the universes in the ensemble.

Moreover, in contrast to parallel worlds, other O-regions are undeniably real. They all belong to the same spacetime, and given enough time, we may even be able to travel to other O-regions and to compare their histories with ours.‡

SOME WAYS OUT

Many readers are, no doubt, wondering, Do we really have to believe all this nonsense about our clones? Is there any way to avoid these bizarre conclusions? If you absolutely cannot stand the thought of your double in a distant galaxy being a Republican (or a Democrat, as the case may be), and if you are willing to clutch at any straw to avoid it, let me offer you a couple of straws.

First of all, there is always a chance that the theory of inflation is wrong. The idea of inflation is very compelling, and the observational signs are en-

---

*This latter view is close to the Copenhagen picture, except it does not insist on the presence of external observers.

†We shall see later, in Chapter 17, that there may in fact be a good reason to believe in the existence of other, completely disconnected universes.

‡Our ability to travel to other O-regions may be hindered if the observed accelerated expansion of the universe is due to a constant vacuum energy. In this case, galaxies in other O-regions will continue moving away faster and faster, and we will never be able to catch up with them. Some models, however, predict that the vacuum energy will gradually subside, as it did during inflation. Then there is no limit, in principle, to how far we can travel.

couraging, but inflation is not nearly as well established as, for example, Einstein's theory of relativity.

Even if our universe is a product of inflation, it is conceivable that inflation is not eternal. This outcome, however, can be achieved only at the cost of making the theory rather contrived. In order to avoid eternal inflation, the energy landscape of the scalar field needs to be custom-tailored specifically for that purpose.[8]

Neither of these options appears to be attractive. The theory of inflation is by far the best explanation we have for the big bang. If we accept this theory, and refuse to mutilate it by adding any ad hoc, unnecessary features, then we have no choice but to accept eternal inflation—with all of its consequences, whether we like them or not.

A FAREWELL TO UNIQUENESS

In times of antiquity we, humans, were at the center of the universe. The sky was not far off, and the fates of kingdoms and individuals could be read from the pattern of stars and planets on its velvet vault. Our descent from center stage started with Copernicus, and by the end of the last century it was nearly complete. Not only is the Earth not the center of the solar system, but the Sun itself is an unremarkable star at the outskirts of a rather typical galaxy. And yet, we could still hold on to the idea that there was something distinctly special about our Earth—that it was the only planet with this particular set of life forms, and that our human civilization, with its art, culture, and history, was unique in the entire universe. One might think that that alone was reason enough to treasure our little planet like a precious work of art.

Now, we have been robbed of this last claim to uniqueness. In the worldview that has emerged from eternal inflation, our Earth and our civilization are anything but unique. Instead, countless identical civilizations are scattered in the infinite expanse of the cosmos. With humankind reduced to absolute cosmic insignificance, our descent from the center of the universe is now complete.[9]

PART III

. . .

PRINCIPLE OF
MEDIOCRITY

12

The Cosmological Constant Problem

*Few theoretical estimates in the history of physics . . . have ever
been so inaccurate.*　　　　　　　　　　　　　—LARRY ABBOTT

VACUUM ENERGY CRISIS

The most mysterious object ever encountered by physicists is the vacuum. And the most daunting secret of the vacuum is the origin of its energy. I should clarify that I do not mean the high-energy false vacuum of inflationary cosmology. False-vacuum physics is, in fact, relatively well understood. The enigmatic object I am talking about is the ordinary, true vacuum that we now inhabit.

Vacuum is what you get when you remove all particles and radiation. For a classical physicist, it is just empty space, and there is not much more to say about it. But in quantum physics, the vacuum is a scene of frenetic activity.

Take, for example, electromagnetic radiation. It consists of photons—little lumps of electromagnetic energy. Suppose you have a box of pure vacuum. You clear the interior of the box and make sure there is not a single photon, or any other particle, left inside. You might think that the electric and magnetic fields in the box should now be strictly equal to zero. But they are not. The quantum vacuum refuses to stay still. Just like the scalar field during inflation, electric and magnetic fields experience random jerks, or quantum fluctuations.

If you try to measure, say, the magnetic field inside the box, the answer you get depends on the size of your measuring device. Suppose you start with a fairly large device, which probes the field on the scale of 1 centimeter. The magnitude of the field will then come out to be a few billionths of a gauss. (To put this in perspective, note that the strength of the magnetic field on the Earth's surface is about 1 gauss.) One nanosecond* later, the direction of the field will be completely different, while its magnitude will be anywhere between zero and a few billionths of a gauss. To detect these rapid fluctuations of the field, you will have to measure it very quickly. If the measurement takes longer than a nanosecond, you will get the averaged value of the field, which is very close to zero.

A 1-millimeter detector would measure a magnetic field that is 100 times stronger and fluctuates 10 times faster. The same pattern holds as you go to still smaller scales: every time you reduce the distance scale by 10, the magnitude of the fluctuations grows by a factor of 100 and their frequency increases tenfold. On the atomic scale, the fluctuating magnetic field is 10 million gauss and changes its direction about 10^{17} times per second.

The reason we are not aware of these huge magnetic fields is that they vary so rapidly from one point to another and from one moment to the next. A compass needle, for example, reacts to the magnetic field averaged over the needle's length and over the time it takes to turn by a noticeable amount (say, 0.1 second). The effect of quantum fluctuations on such scales is completely negligible.[1]

All is well until we check the energy of the fluctuations. The energy density in a magnetic field depends only on the field strength, not on its direction. Hence, even though the magnetic field fluctuates back and forth, its energy density does not average to zero. Large, rapidly fluctuating fields on smaller-distance scales give a greater contribution to the energy density. And that's where we run into a problem. As we include fluctuations on smaller and smaller scales, the energy density grows without bound. Thus, we arrive at the absurd conclusion that the energy density of the vacuum is infi-

---

*A nanosecond is one-billionth of a second.

nite! Something clearly went very wrong with our theory. Let us try to see what this could be and how we can avoid this bizarre conclusion.

The infinity arises when we allow the length scale of the fluctuations to get arbitrarily small. But there may be a limit to how small it can be. At supersmall distances, the geometry of space and time is also subject to large quantum fluctuations. As in the case of electromagnetism, the smaller the distance scale, the larger the fluctuations. Below a certain critical distance, called the *Planck length*, spacetime acquires a chaotic, foamlike structure. The space warps and twists violently, small disconnected space "bubbles" pop out and collapse, and multiple "handles," or "tunnels," are being created and instantly destroyed (see Figure 12.1). The Planck length is incredibly small: it is one billion-trillion-trillionth of a centimeter. On much larger scales, the space appears to be smooth and the "spacetime foam" is not visible—just as the foamy surface of the ocean appears smooth when viewed from a large distance.

It is possible that the drastic change in the character of spacetime intervenes to suppress the runaway electromagnetic fluctuations. We cannot tell for sure, since the physics of spacetime foam is not well understood. But even in the best-case scenario, there seems to be nothing to restrain the fluctuations on scales greater than the Planck length. An estimate of the energy density of such fluctuations gives an astounding 10^{88} tons per cubic centimeter, much higher than in the grand-unified vacuum!

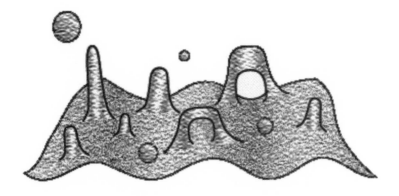

Figure 12.1. Spacetime foam.

The energy density of the true vacuum is what Einstein called the cosmological constant. If indeed it were so tremendously large, the universe would now be in the state of explosive inflationary expansion. But the observed rate of expansion of the universe puts a bound on the cosmological constant, which is 10^{120} (more than a google!) times smaller. We thus have a puzzle on our hands: Why isn't the vacuum energy density huge? The glaring discrepancy between the predicted and observed values of the cosmological constant is known as the cosmological constant problem. It is one of the most tantalizing and frustrating mysteries that we now face in theoretical particle physics.

IN SEARCH OF A DEEP SYMMETRY

Apart from electromagnetism, quantum fluctuations of other fields also contribute to the vacuum energy. It turns out that some of these contributions are negative, and there is some hope that positive and negative energy contributions might compensate one another. This possibility has inspired numerous attempts to solve the cosmological constant problem.

All elementary particles can be divided into two types: *bosons* and *fermions*.* Photons, for example, are bosons, and electrons, positrons, and quarks are fermions. Fermi-particles can be pictured as small bundles of fermionic fields, but in contrast to electromagnetism, the magnitudes of such fields are characterized by the Grassmann numbers,† which are very different from ordinary numbers. When you multiply ordinary numbers, the result does not depend on the order of multiplication; for instance, $4 \times 6 = 6 \times 4 = 24$. But for Grassmann numbers the product changes sign if you reverse the order of multiplication: $a \times b = -b \times a$. The Grassmann character of the fermionic fields is responsible for many distinctive features of Fermi-particles, but what is important for us here is that the vacuum fluctuations of Fermi fields have a negative energy density.

Could it be that the positive vacuum energy of Bose fields is compensated by the negative energy of Fermi fields? This is possible in principle,

---

*So named after Satyendra Bose and Enrico Fermi, who elucidated their distinctive properties.

†Named after the nineteenth-century German mathematician Hermann Grassmann, who first introduced them.

but seems extremely unlikely. The huge positive and negative terms, which depend in complicated ways on particle masses and interactions, have to cancel one another with an accuracy better than one part in a google. What could have caused such a miraculous coincidence?

Remarkable cancellations do occur in particle physics, but they can usually be traced to some underlying symmetry. Take, for example, electric charge conservation. A high-energy collision can produce myriads of new particles, but you can always be sure that the numbers of positively and negatively charged particles that have been created are exactly equal, so the total charge is unchanged. This property is due to a special symmetry of the equations of elementary particle physics, called *gauge symmetry*.*

It follows from gauge symmetry that electric charge is conserved in all elementary particle interactions. The beauty of symmetry is that details are unimportant. It does not matter what particle masses are or what kinds of interactions they are involved in. Charge conservation follows anyway.

Until very recently, the great majority of physicists believed that something of this sort should be going on in the case of the vacuum energy. There should be some deep symmetry, waiting to be discovered, that enforces the cancellation of different contributions to the cosmological constant.[2] Since the 1970s, numerous attempts have been made to figure out what this symmetry might be—some of them by the best minds in theoretical physics. However, after several decades, there was little to show for all that effort. The cosmological constant problem looked as formidable as ever.

THE COINCIDENCE PROBLEM

> *"Any coincidence," said Miss Marple to herself, "is worth noticing. You can throw it away later if it is only a coincidence."*
> —AGATHA CHRISTIE, *Nemesis*

It came as a total surprise when in the late 1990s two teams of astronomers announced that they had evidence for a nonvanishing cosmological constant. As we discussed in Chapter 9, this discovery was great news for the

---

*An equation is said to have symmetry if there is some operation that leaves it unchanged. For example, the equation $x + y = 1$ does not change if we swap x and y.

theory of inflation. The mass (energy) density of the vacuum provided precisely the amount that was missing to make the universe flat. But it was dreadful news for the particle theory.

The goal of solving the cosmological constant problem with a beautiful symmetry appeared now even more elusive. A symmetry would do a perfect job; it would not leave even a trace of vacuum energy uncompensated. But that was not all. The actual value of the cosmological constant that was obtained from the data looked extremely suspicious—so much so that most particle physicists and cosmologists refused to believe it and hoped that it would somehow go away.

The observed mass density of the vacuum is slightly more than twice the average density of matter. The puzzle is that the two densities are comparable, in the sense that one is not very much greater or smaller than the other. This is surprising, because the matter density and the vacuum density behave very differently with the expansion of the universe. The vacuum density does not change at all (as long as we stay in the same vacuum), while the matter density decreases as the volume grows. If the two densities are more or less the same today, then at the time of last scattering the matter density was a billion times greater than the vacuum density, and at 1 second A.B. it was 10^{45} times greater. In the distant future the pattern will be reversed and the density of matter will become much smaller than that of the vacuum. For example, a trillion years from now it will be 10^{50} times smaller.

Thus, throughout most of the history of the universe the density of matter is strikingly different from that of the vacuum. Why, then, do we happen to live at the very special epoch when the two densities are close to each other? Considering the huge range of variation of the matter density, the coincidence is so extraordinary that it's very hard to dismiss it as "only a coincidence."

It looked as if nature were trying to tell us something. But, in her usual manner, she refused to make it easy for us to understand. Why would a fundamental constant of nature, like the cosmological constant, be related to the matter density at the particular epoch when we humans happen to be around? The idea of some connection between these two quantities appeared totally ridiculous. The particle physics community was in disarray.

And then there was a remarkable fact that made the situation even more peculiar. A nonzero cosmological constant, not far off the observed value, had been theoretically predicted years before the observations were made. But there was a problem with that prediction. It was based on *anthropic selection*—an idea so controversial that most self-respecting physicists avoided it like the plague.

13

Anthropic Feuds

We have found a strange footprint on the shores of the unknown. We have devised profound theories, one after another, to account for its origin. At last, we have succeeded in reconstructing the creature that made the footprint. And Lo! It is our own.

—Sir Arthur Eddington

CONSTANTS OF NATURE

The properties of every object in the universe, from a DNA molecule to a giant galaxy, are determined, in the final analysis, by several numbers—the constants of nature. These constants include the masses of elementary particles and the parameters characterizing the strength of the four basic interactions, or forces—*strong, weak, electromagnetic,* and *gravitational.* The proton, for example, is 0.14 percent less massive than the neutron and 1836 times more massive than the electron.* The gravitational attraction between two protons is 10^{40} times weaker than their electric repulsion. On the face of it, these numbers appear completely arbitrary. To borrow Craig Hogan's metaphor,[1] we can imagine the Creator sit-

---

*The numerical value of the mass depends on the units used to measure it (e.g., grams, ounces, or atomic units), but a ratio of two masses, like 1836, is independent of this choice.

Figure 13.1. At the control board of the universe.

ting at the control board of the universe and turning different knobs to adjust the values of the constants. "Shall we make it 1835 or 1836?"

Or could it be that there is some system behind this seemingly random set of numbers? Maybe there are no knobs to twiddle and the numbers are all fixed by mathematical necessity. It has long been a dream of particle physicists that indeed there is no choice and that all constants of nature will eventually be derived from some yet-to-be-discovered fundamental theory.

As of now, however, we have no indication that the choice of the constants is preordained. The Standard Model of particle physics, which describes strong, weak, and electromagnetic interactions of all known particles, contains twenty-five "adjustable" constants. The values of these constants are determined from observations.* Together with the newly discovered cosmological constant, we thus need twenty-six constants of nature to de-

---

*The values of some of these constants, particularly those characterizing the properties of neutrinos, are still unknown.

scribe the physical world. The list may have to be extended if new particles or new types of interaction are discovered.

FINE-TUNING THE UNIVERSE

The Creator's choice of the constants may appear rather capricious, and yet, remarkably, there does seem to be a system behind it—although not of the kind physicists have been hoping for. Research in diverse areas of physics has shown that many essential features of the universe are sensitive to the precise values of some of the constants. Had the Creator adjusted the knobs slightly differently, the universe would be a strikingly different place. And most likely neither we, nor any other living creatures, would be around to admire it.

To start with, let us consider the effect of varying the neutron mass. As it stands now, it is slightly greater than the proton mass, which allows free neutrons to decay into protons and electrons.* Suppose now that we turn the neutron mass knob toward smaller values. It takes a very small change, no more than 0.2 percent, for the mass difference between proton and neutron to reverse. Now protons become unstable and decay into neutrons and positrons. Protons may still be stabilized inside atomic nuclei, but with some further turning of the knob they will decay there as well. As a result, the nuclei will lose their electric charge and atoms will disintegrate, since there will be nothing to keep electrons in orbit around the nuclei. The unattached electrons will form close pairs with the positrons. They will swirl around one another in a deadly dance and quickly annihilate into photons. We will thus be left in a "neutron world," consisting of isolated neutronic nuclei and radiation. This world has no chemistry, no complex structures, and no life.

We next turn the neutron mass knob in the opposite direction. Once again, a mass increase of only a fraction of a percent triggers a catastrophic change. As neutrons get heavier, they become more unstable, and at some point they start decaying inside the atomic nuclei, turning into protons. The nuclei are then torn apart by the electric repulsion between protons, and the protons, once they are freed from the nuclei, combine with electrons to

*The decay is accompanied by emission of an antineutrino.

form hydrogen atoms. Thus, we end up in a rather dull "hydrogen world," where no chemical elements can exist except hydrogen.*

To proceed with our exploration, let us now examine the effect of varying the strengths of basic particle interactions. Weak interactions do not play much of a role in the present-day universe, except in spectacular stellar explosions—the supernovae. When a massive star runs out of nuclear fuel, the inner core of the star collapses under its own weight. Enormous energy is released, escaping mostly in the form of weakly interacting neutrinos. Photons and other particles, which interact strongly or electromagnetically, remain trapped in the superdense collapsing core. On their way out, neutrinos blow off the outer layers of the star, which results in a colossal explosion. If weak interactions were much stronger than they actually are, neutrinos would not be able to escape from the core, and if they were much weaker, neutrinos would fly freely through outer layers without dragging them along. Thus, if we were to make a significant change in the strength of weak interactions one way or the other, astronomers would lose one of their most cherished spectacles.

You think you might be able to live with that? But wait; let us not turn the knob just yet. The effect of the change at earlier stages of cosmic evolution could be much more devastating. As we discussed in Chapter 4, heavy elements, such as carbon, oxygen, and iron, were forged in stellar interiors and then dispersed in supernova explosions. These elements are essential for the formation of planets and living creatures. Without supernovae, they would remain buried inside stars and the only elements available would be the lightest ones, formed in the big bang: hydrogen, helium, and deuterium, with a trace of lithium—not the kind of universe you would like to live in.

Gravity is by far the weakest of the four fundamental forces. Its effects are important only in the presence of huge aggregates of matter, like galaxies or stars. In fact, it is the weakness of gravity that makes the stars so massive: the mass has to be large enough to squeeze the hot gas to the high

---

*On a more fundamental level, protons and neutrons are made up of quarks, so it is more appropriate to regard their masses as derived quantities and the quark masses as fundamental constants of nature. This, however, does not change the general conclusions. A few percentage points' variation of the quark masses drives us either into a neutron world or into a hydrogen world.

density needed for nuclear reactions. If we were to make gravity stronger, the stars would not be so large and would burn out faster. A millionfold increase in the strength of gravity would make stellar masses a billion times smaller.* The mass of a typical star would then be less than the present mass of the Moon, and its lifetime would be about 10,000 years (compared to 10 billion years for the Sun). This time interval is hardly long enough for even the simplest bacteria to evolve. A much smaller enhancement of gravity may in fact be sufficient to depopulate the universe. A hundredfold increase, for example, would reduce stellar lifetimes well below the few billion years that it took for intelligent life to evolve on Earth.

These and many other examples show that our presence in the universe depends on a precarious balance between different tendencies—a balance that would be destroyed if the constants of nature were to deviate significantly from their present values.[2] What are we to make of this fine-tuning of the constants? Is it a sign of a Creator who carefully adjusted the constants so that life and intelligence would be possible? Perhaps. But there is also a completely different explanation.

THE ANTHROPIC PRINCIPLE

The alternative view is based on a very different image of the Creator. Instead of meticulously designing the universe, he botches one sloppy job after another, putting out a huge number of universes with different and totally random values of the constants. Most of these universes are as exciting as the neutron world, but once in a while, by pure chance, a finely tuned universe fit for life will be created.

Given this worldview, let us ask ourselves: What kind of universe can we expect to live in? Most of the universes will be dreary and unsuitable for life, but there will be nobody there to complain about that. All intelligent beings will find themselves in the rare bio-friendly universes and will marvel at the miraculous conspiracy of the constants that made their existence possible. This line of reasoning is known as the *anthropic principle*. The name

---

*Note that even after a millionfold enhancement, gravity would still be 10^{34} times weaker than electromagnetism.

was coined in 1974 by Cambridge astrophysicist Brandon Carter,* who offered the following formulation of the principle: "[W]hat we can expect to observe must be restricted by the conditions necessary for our presence as observers."[3]

The anthropic principle is a selection criterion. It assumes the existence of some distant domains where the constants of nature are different. These domains may be located in some remote parts of our own universe, or they could belong to other, completely disconnected spacetimes. A collection of domains with a wide variety of properties is called a *multiverse*—the term introduced by Carter's former classmate Martin Rees, now Britain's Astronomer Royal. Later in this book we shall encounter three types of multiverse ensembles. The first consists of a multitude of regions all belonging to the same universe. The second type is made up of separate, disconnected universes.† And the third type is a combination of the two: it consists of multiple universes, each of which has a variety of different regions. If a multiverse of any type really exists, then it is not surprising that the constants of nature are fine-tuned for life. On the contrary, they are guaranteed to be fine-tuned.

Anthropic reasoning can also be applied to variations of observable properties in time, rather than in space. One of the earliest applications was by Robert Dicke, who used the anthropic approach to explain the present age of the universe. Dicke argued that life can form only after heavy elements are synthesized in stellar interiors. This takes a few billion years. The elements are then dispersed in supernova explosions, and we have to allow a few more billion years for the second generation of stars and their planetary systems to form in the aftermath of the explosions and for biological evolution to occur. The first observers could not, therefore, appear much earlier than 10 billion years A.B. We should also keep in mind that a star like our Sun exhausts its nuclear energy in about 10 billion years and that the galactic supply of gas for new star formation is also depleted on a similar

---

*Now at Meudon Observatory in France.

†Philosophers often define the universe as "everything there is." Then, of course, there cannot be any other universes. Physicists do not usually use the term in this broadest sense and refer to completely disjointed, self-contained spacetimes as separate universes. Here I follow the physics tradition.

time scale. At 100 billion years A.B. there will be very few Sun-like stars left in the visible universe.[4] If we assume that life will perish with the death of stars, we are left with a window between, say, 5 and 100 billion years A.B. when observers can exist.* Not surprisingly, the present age of the universe falls within this window.[5]

Dicke's use of the anthropic principle to constrain our location in time was uncontroversial. But Brandon Carter, Martin Rees, and a few other physicists attempted to go beyond that, using anthropic reasoning to explain the fine-tuning of the fundamental constants. And that's where the controversy began.

WHAT DOES THE ANTHROPIC PRINCIPLE HAVE IN COMMON WITH PORNOGRAPHY?

As formulated by Carter, the anthropic principle is trivially true. The constants of nature and our location in spacetime should not preclude the existence of observers. For otherwise our theories would be logically inconsistent. When interpreted in this sense, as a simple consistency requirement, the anthropic principle is, of course, uncontroversial, although not very useful. But any attempt to use it as an *explanation* for the fine-tuning of the universe evoked an adverse and unusually temperamental response from the physics community.

There were in fact some good reasons for that. In order to explain the fine-tuning, one has to postulate the existence of a multiverse, consisting of remote domains where the constants of nature are different. The problem is, however, that there is not one iota of evidence to support this hypothesis. Even worse, it does not seem possible to *ever* confirm or disprove it. The philosopher Karl Popper has argued that any statement that cannot be falsified cannot be scientific. This criterion, which has been generally adopted by physicists, seems to imply that anthropic explanations of the fine-tuning are not scientific. Another, related criticism was that the anthropic principle can only be used to explain what we already know. It never predicts anything, and thus cannot be tested.

---

*It is conceivable that advanced civilizations can survive the death of stars using nuclear energy or the energy of tides to sustain life. But it appears more likely that civilizations are relatively short-lived. I will touch upon this subject in Chapter 14.

It did not help that the whole subject of the anthropic principle had been obscured by murky and confusing interpretations.* On top of that, many different formulations of the principle appeared in the literature (the philosopher Nick Bostrom, who wrote a book on the subject,[6] counted more than thirty). The situation is well summarized by a quote from Mark Twain: "The researches of many commentators have already thrown much darkness on this subject, and it is probable that, if they continue, we shall soon know nothing at all about it."[7] The term "anthropic" was itself a source of confusion, as it seems to refer to human beings, rather than to intelligent observers in general.

But the main reason why the response to anthropic explanations was so emotional was probably the feeling of betrayal. Ever since Einstein, physicists believed that the day will come when all constants of nature will be calculated from some all-encompassing Theory of Everything. Resorting to anthropic arguments was viewed as a capitulation and evoked reactions ranging from annoyance to outright hostility. Some well-known physicists went so far as to say that anthropic ideas were "dangerous"[8] and that they were "corrupting science."[9] Only in extreme cases, when all other possibilities have been exhausted, might one be excused for mentioning the "A-word," and sometimes not even then. The Nobel Prize winner Steven Weinberg once said that a physicist talking about the anthropic principle "runs the same kind of risk as a cleric talking about pornography. No matter how much you say you are against it, some people will think you are a little too interested."

THE COSMOLOGICAL CONSTANT

If there has ever been a problem calling for measures of last resort, it is the cosmological constant problem. Different contributions to the vacuum energy density conspire to cancel one another with an accuracy of one part in 10^{120}. This is the most notorious and perplexing case of fine-tuning in

---

*Carter himself contributed to the confusion by introducing an alternative version of the principle, called the "strong anthropic principle," which states that "the universe . . . must be such as to admit the creation of observers within it at some stage." Many people interpreted this in a mystical sense, as referring to some sort of theological necessity. In this book I adopt Carter's original formulation, which he referred to as the "weak anthropic principle."

physics. Andrei Linde was one of the first brave souls to apply anthropic reasoning to this problem. He was not satisfied with the vague talk about "other universes" and suggested a specific model of how the cosmological constant could be variable and what could make it change from one place to another.

Linde used an idea that had worked for him before. Remember the little ball rolling down in the energy landscape? The ball represented a scalar field; and its elevation, the energy density of the field. As the field rolled downhill, its energy drove the inflationary expansion of the universe.

The feature Linde took from this model of inflation was that different elevations in the landscape correspond to different energy densities. He assumed the existence of another scalar field with an energy landscape of its own. To avoid confusion with the field responsible for inflation, we shall call the latter field "the inflaton"—its usual name in the physics literature. In our neighborhood, the inflaton has already rolled to the bottom of its energy hill. (This happened 14 billion years ago, at the end of inflation.) To prevent his new field from rolling downhill too quickly, Linde had to require that the slope should be exceedingly gentle, much more so than in the model of inflation. Any slope, no matter how small, will eventually cause the field to roll down. With smaller slopes it will take longer "to get the ball rolling." Linde assumed the slope to be so small that the field would not move much in the 14 billion years that elapsed since the big bang. But if the slope extends to a great length in both directions, the energy density can reach very large positive or negative values. (See Figure 13.2.)

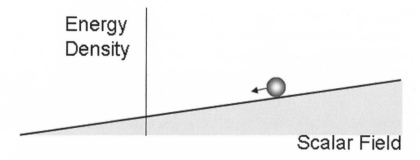

Figure 13.2. A scalar field on a very gentle slope of the energy landscape.

The full energy density of the vacuum—the cosmological constant—is obtained by adding the energy density of the scalar field to the vacuum energy densities of fermions and bosons calculated from particle physics. Even if there are no miraculous cancellations and the particle physics contribution is huge, there will be a spot on the slope where the scalar field contribution has an equal magnitude and opposite sign, so the total vacuum energy density is zero. The scalar field is presumably very close to that spot in our part of the universe.

If the scalar field were to vary from one part of the universe to another, the cosmological "constant" would also be variable, and that is all one needs to apply the anthropic principle. But what could cause the scalar field to vary? Linde had a good answer to this one as well!

Prior to the big bang, in the course of eternal inflation, the field experienced random quantum kicks. As before, we can represent the behavior of the field by that of a party of random walkers (see Chapter 8). The slope of the hill is too small to be of any importance in this case, so the walkers step left and right with nearly equal probability. Even if they start at the same place, the walkers will gradually drift apart and, given enough time, will spread along the entire slope. (Recall, there is no shortage of time in eternal inflation.) Since the walkers represent scalar field values in different regions of space, we conclude that quantum processes during inflation necessarily generate a distribution of regions with all possible values of the field—and therefore all possible values of the cosmological constant.

While the walkers are wandering on the slope, the distances between the regions they represent are being stretched by the exponential inflationary expansion. As a result, the spatial variation of the vacuum energy density is extremely small.* You would have to travel googles of miles before you noticed the slightest change.

Linde's model can be extended to include more scalar fields and make other constants of nature variable.† And if the fundamental particle physics allows the constants to vary, then quantum processes during eternal inflation inescapably generate vast regions of space with all possible values of

---

*In order to get to a noticeably different elevation, a random walker would have to travel a long distance along the very flat slope. In the meantime, the universe would expand by a huge amount.

†It is not clear whether or not scalar fields of the kind postulated by Linde really exist. We shall return to this issue in Chapter 15.

the constants. Eternal inflation thus provides a natural arena for applications of the anthropic principle.

Now that we have an ensemble of regions with different values of the cosmological constant, what value should we expect to observe? In regions where the mass density of the vacuum is greater than the density of water (1 gram per cubic centimeter), stars would be torn apart by repulsive gravity. It turns out, however, that a much smaller vacuum density would do enough damage to make observers impossible. This was shown by Steven Weinberg, in a paper that later became a classic of anthropic reasoning.

As the universe expands, the density of matter is diluted, and inevitably there comes a time when it drops below that of the vacuum. Weinberg found that once this happens, matter can no longer clump into galaxies; instead it is dispersed by the repulsive vacuum gravity. The larger the cosmological constant is, the earlier is the time of vacuum dominance. And regions

Figure 13.3. Steven Weinberg. (Photo by Frank Curry, Studio Penumbra)

where it dominates before any galaxies have had a chance to form will have no cosmologists to worry about the cosmological constant problem.

The effect of a negative cosmological constant is even more devastating. In this case, the vacuum gravity is attractive, and vacuum domination leads to a rapid contraction and collapse of the corresponding regions. The anthropic principle requires that the collapse should not occur before galaxies and observers have had time to evolve.

According to Weinberg's analysis, the largest mass density of the vacuum that still allows some galaxies to form is about the mass of a few hundred hydrogen atoms per cubic meter—10^{27} times smaller than the density of water. This was a great improvement over the googles of tons per cubic centimeter suggested by particle physicists' calculations.

If indeed the smallness of the cosmological constant is due to anthropic selection, then, small though it is, the constant does not have to be exactly zero. In fact, there seems to be no reason why it should be much smaller than required by the anthropic principle. In the late 1980s, the observational accuracy was already reaching the level necessary to detect such values of the constant, and Weinberg made a prediction that it would soon show up in astronomical observations. Indeed, almost a decade later the first hints of cosmological constant appeared in supernova data.

14

Mediocrity Raised to a Principle

*I consider myself an average man, except for the fact that I con-
sider myself an average man.* —MICHEL DE MONTAIGNE

THE BELL CURVE

The most scathing criticism raised against the anthropic principle is
that it does not yield any testable predictions. All it says is that we can
observe only those values of the constants that allow observers to exist. This can hardly be regarded as a prediction, since it is guaranteed to be
true. The question is, Can we do any better? Is it possible to extract some
nontrivial predictions from anthropic arguments?

If the quantity I am going to measure can take a range of values, determined largely by chance, then I cannot predict the result of the measurement with certainty. But I can still try to make a statistical prediction.
Suppose, for example, I want to predict the height of the first man I am going to see when I walk out into the street. According to the *Guinness Book of
Records*, the tallest man in medical history was the American Robert Pershing Wadlow, whose height was 2.72 meters (8 feet 11 inches). The shortest
adult man, the Indian Gul Mohammed, was just 56 centimeters tall (about
22 inches). If I want to play it really safe, I should predict that the first man
I see will be somewhere between these two extremes. Barring the possibil-

ity of breaking the Guinness records, this prediction is guaranteed to be correct.

To make a more meaningful prediction, I could consult the statistical data on the height of men in the United States. The height distribution follows a bell curve, shown in Figure 14.1, with a median value at 1.77 meters (about 5 feet 9½ inches). (That is, 50 percent of men are shorter and 50 percent are taller than this value.) The first man I meet is not likely to be a giant or a dwarf, so I expect his height to be in the mid-range of the distribution. To make the prediction more quantitative, I can assume that he will not be among the tallest 2.5 percent or shortest 2.5 percent of men in the United States. The remaining 95 percent have heights between 1.63 meters (5 feet 4 inches) and 1.90 meters (6 feet 3 inches). If I predict that the man I meet will be within this range of heights and then perform the experiment a large number of times, I can expect to be right 95 percent of the time. This is known as a prediction at 95 percent confidence level.

In order to make a 99 percent confidence level prediction, I would have to discard 0.5 percent at both ends of the distribution. As the confidence level is increased, my chances of being wrong get smaller, but the predicted range of heights gets wider and the prediction less interesting.

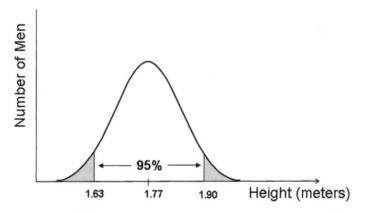

Figure 14.1. Height distribution of men in the United States. The number of men whose height is within a given interval is proportional to the area under the corresponding portion of the curve. The shaded "tails" of the bell curve mark 2.5 percent at low and high ends of the distribution. The range between the marked areas is predicted at 95 percent confidence level.

Can a similar technique be applied to make predictions for the constants of nature? I was trying to find the answer to this question in the summer of 1994, when I visited my friend Thibault Damour at the Institut des Hautes Études Scientifiques in France. The institute is located in a small village, Bures-sur-Yvette, a thirty-minute train ride from Paris. I love the French countryside and, despite the calories, French food and wine. The famous Russian physicist Lev Landau used to say that a single alcoholic drink was enough to kill his inspiration for a week. Luckily, this has not been my experience. In the evenings, with my spirits up after a very enjoyable dinner, I would take a walk in the meadows along the little river Yvette, and my thoughts would gradually return to the problem of anthropic predictions.

Suppose some constant of nature, call it X, varies from one region of the universe to another. In some of the regions observers are disallowed, while in others observers can exist and the value of X will be measured. Suppose further that some Statistical Bureau of the Universe collected and published the results of these measurements. The distribution of values measured by different observers would most likely have the shape of a bell curve, similar to the one in Figure 14.1. We could then discard 2.5 percent at both ends of the distribution and predict the value of X at a 95 percent confidence level.

Figure 14.2. An observer randomly picked in the universe. The values of the constants measured by this observer can be predicted from a statistical distribution.

What would be the meaning of such a prediction? If we randomly picked observers in the universe, their observed values of X would be in the predicted interval 95 percent of the time. Unfortunately, we cannot test this kind of prediction, because all regions with different values of X are beyond our horizon. We can only measure X in our local region. What we can do, though, is to think of ourselves as having been randomly picked. We are just one out of a multitude of civilizations scattered throughout the universe. We have no reason to believe a priori that the value of X in our region is very rare, or otherwise very special compared with the values measured by other observers. Hence, we can predict, at 95 percent confidence level, that our measurements will yield a value in the specified range. The assumption of being unexceptional is crucial in this approach; I called it "the principle of mediocrity."

Some of my colleagues objected to this name. They suggested "the principle of democracy" instead. Of course, nobody wants to be mediocre, but the name expresses nostalgia for the times when humans were at the center of the world. It is tempting to believe that we are special, but in cosmology, time and again, the assumption of being mediocre proved to be a very fruitful hypothesis.

The same kind of reasoning can be applied to predicting the height of people. Imagine for a moment that you don't know your own height. Then you can use statistical data for your country and gender to predict it. If, for example, you are an adult man living in the United States and have no reason to think that you are unusually tall or short, you can expect, at 95 percent confidence, to be between 1.63 and 1.90 meters tall.

I later learned that similar ideas had been suggested earlier by the philosopher John Leslie and, independently, by the Princeton astrophysicist Richard Gott. The main interest of these authors was in predicting the longevity of the human race. They argued that humanity is not likely to last much longer than it has already existed, since otherwise we would find ourselves to be born surprisingly early in its history. This is what's called the "doomsday argument." It dates back to Brandon Carter, the inventor of the anthropic principle, who presented the argument in a 1983 lecture, but never in print (it appears that Carter already had enough controversy on his hands).[1] Gott also used a similar argument to predict the fall of the Berlin Wall and the lifetime of the British journal *Nature*, where he published his first paper on this topic. (The latter prediction, that *Nature* will go out of print by the year 6800, is yet to be verified.)

If we have a statistical distribution for the constants of nature measured by all the observers in the universe, we can use the principle of mediocrity for making predictions at a specified confidence level. But where are we going to get the distribution? In lieu of the data from the Statistical Bureau of the Universe, we will have to derive it from theoretical calculations.

The statistical distribution cannot be found without a theory describing the multiverse with variable constants. At present, our best candidate for such a theory is the theory of eternal inflation. As we discussed in the preceding chapter, quantum processes in the inflating spacetime spawn a multitude of domains with all possible values of the constants. We can try to calculate the distribution for the constants from the theory of eternal inflation, and then—perhaps!—we could check the results against the experimental data. This opens an exciting possibility that eternal inflation can, after all, be subjected to observational tests. Of course, I felt this opportunity was not to be missed.

COUNTING OBSERVERS

Consider a large volume of space, so large that it includes regions with all possible values of the constants. Some of these regions are densely populated with intelligent observers. Other regions, less favorable to life, are greater in volume, but more sparsely populated. Most of the volume will be occupied by huge barren domains, where observers cannot exist.

The number of observers who will measure certain values of the constants is determined by two factors: the volume of those regions where the constants have the specified values (in cubic light-years, for example), and the number of observers per cubic light-year. The volume factor can be calculated from the theory of inflation, combined with a particle physics model for variable constants (like the scalar field model for the cosmological constant).[2] But the second factor, the population density of observers, is much more problematic. We know very little about the origin of life, let alone intelligence. How, then, can we hope to calculate the number of observers?

What comes to the rescue is that some of the constants do not directly affect the physics and chemistry of life. Examples are the cosmological constant, the neutrino mass, and the parameter, usually denoted by Q, that characterizes the magnitude of primordial density perturbations. Variation of such life-neutral constants may influence the formation of galaxies, but not the chances for life to evolve within a given galaxy. In contrast, constants such as the electron mass or Newton's gravitational constant have a direct impact on life processes. Our ignorance about life and intelligence can be factored out if we focus on those regions where the life-altering constants have the same values as in our neighborhood and only the life-neutral constants are different. All galaxies in such regions will have about the same number of observers, so the density of observers will simply be proportional to the density of galaxies.[3]

Thus, the strategy is to restrict the analysis to life-neutral constants. The problem then reduces to the calculation of how many galaxies will form per given volume of space—a well-studied astrophysical problem. The result of this calculation, together with the volume factor derived from the theory of inflation, will yield the statistical distribution we are looking for.

CONVERGING ON THE COSMOLOGICAL CONSTANT

As I was thinking about observers in remote domains with different constants of nature, it was hard to believe that the equations I scribbled in my notepad had much to do with reality. But having gone this far, I bravely pressed ahead: I wanted to see if the principle of mediocrity could shed some new light on the cosmological constant problem.

The first step was already taken by Steven Weinberg. He studied how the cosmological constant affects galaxy formation and found the anthropic bound on the constant—the value above which the vacuum energy would dominate the universe too soon for any galaxies to form. Moreover, as I already mentioned, Weinberg realized that there was a *prediction* implicit in his analysis. If you pick a random value between zero and the anthropic bound, this value is not likely to be much smaller than the bound, for the same reason that the first man you meet is not likely to be a dwarf. Weinberg argued, therefore, that the cosmological constant in our part of the universe should be comparable to the anthropic bound.*

The argument sounded convincing, but I had my reservations. In regions where the cosmological constant is comparable to the anthropic bound, galaxy formation is barely possible and the density of observers is very low. Most observers are to be found in regions teeming with galaxies, where the cosmological constant is well below the bound—small enough to dominate the universe only after the process of galaxy formation is more or less complete. The principle of mediocrity says that we are most likely to find ourselves among these observers.

I made a rough estimate, which suggested that the cosmological constant measured by a typical observer should not be much greater than ten times the average density of matter. A much smaller value is also improbable—like the chance of meeting a dwarf. I published this analysis in 1995, predicting that we should measure a value of about ten times the matter density in our local region.[4] More detailed calculations, also based on the principle of mediocrity, were later performed by the Oxford astrophysicist George Efstathiou[5] and by Steven Weinberg, who was now joined by his University of Texas colleagues Hugo Martel and Paul Shapiro. They arrived at similar conclusions.

I was very excited about this newly discovered possibility of turning anthropic arguments into testable predictions. But very few people shared my enthusiasm. One of the leading superstring theorists, Joseph Polchinski, once said that he would quit physics if a nonzero cosmological constant were discovered.† Polchinski realized that the only explanation for a small cosmological constant would be the anthropic one, and he just could not stand the thought. My talks about anthropic predictions were sometimes followed by an embarrassed silence. After one of my seminars, a prominent Princeton cosmologist rose from his seat and said, "Anyone who wants to work on the anthropic principle—should." The tone of his remark left little doubt that he believed all such people would be wasting their time.

---

*The anthropic bound derived by Weinberg was somewhat too high for comfort—about 500 times greater than the average density of matter in the universe. In the mid-1990s, observations already indicated that the cosmological constant in our region was at least 50 times smaller. Besides, Weinberg's bound was based on the most distant galaxies known in the late 1980s. By now, even more distant galaxies have been discovered, and the corresponding bound would be 4000 times the average matter density.

†This story was related to me by Sean Carroll of the University of Chicago.

SUPERNOVAE TO THE RESCUE

As I already mentioned in earlier chapters, it came as a complete shock to most physicists when evidence for a nonzero cosmological constant was first announced. The evidence was based on the study of distant supernova explosions of a special kind—type Ia supernovae.

These gigantic explosions are believed to occur in binary stellar systems, consisting of an active star and a white dwarf—a compact remnant of a star that ran out of its nuclear fuel. A solitary white dwarf will slowly fade away, but if it has a companion, it may end its life with fireworks. Some of the gas ejected from the companion star could be captured by the white dwarf, so the mass of the dwarf would steadily grow. There is, however, a maximum mass that a white dwarf can have—the Chandrasekhar limit—beyond which gravity causes it to collapse, igniting a tremendous thermonuclear explosion. This is what we see as a type Ia supernova.

A supernova appears as a brilliant spot in the sky and, at the peak of its brightness, can be as luminous as 4 billion suns. In a galaxy like ours, one type Ia supernova explodes once in about 300 years. So, in order to find a few dozen such explosions, astronomers had to monitor thousands of galaxies over a period of several years. But the effort was worth it. Type Ia supernovae come very close to realizing the long-standing astronomer's dream of finding a *standard candle*—a class of astronomical objects that have exactly the same power. Distances to standard candles can be determined from their apparent brightness—in the same way as we could determine the distance to a 100-watt lightbulb from how bright it appears. Without such magic objects, distance determination is notoriously difficult in astronomy.

Type Ia supernovae have nearly the same power because the exploding white dwarfs have the same mass, equal to the Chandrasekhar limit.[6] Knowing the power, we can find the distance to the supernova, and once we know the distance, it is easy to find the time of the explosion—by just counting back the time it took light to traverse that distance. In addition, the reddening, or Doppler shift, of the light can be used to find how fast the universe was expanding at that time.[7] Thus, by analyzing the light from distant supernovae, we can uncover the history of cosmic expansion.

This technique was perfected by two competing groups of astronomers, one called the Supernova Cosmology Project and the other the High-Redshift Supernova Search Team. The two groups raced to determine the rate at which cosmic expansion was slowed down by gravity. But this was not what they found. In the winter of 1998, the High-Redshift team announced they had convincing evidence that instead of slowing down, the expansion of the universe had been speeding up for the last 5 billion years or so. It took some courage to come out with this claim, since an accelerated expansion was a telltale sign of a cosmological constant. When asked how he felt about this development, one of the leaders of the team, Brian Schmidt, said that his reaction was "somewhere between amazement and horror."[8]

A few months later, the Supernova Cosmology Project team announced very similar conclusions. As the leader of the team, Saul Perlmutter, put it, the results of the two groups were "in violent agreement."

The discovery sent shock waves through the physics community. Some people simply refused to believe the result. Slava Mukhanov* offered me a bet that the evidence for a cosmological constant would soon evaporate. The bet was for a bottle of Bordeaux. When Mukhanov eventually produced the wine, we enjoyed it together; apparently, the presence of the cosmological constant did not affect the bouquet.

There were also suggestions that the brightness of a supernova could be affected by factors other than the distance. For example, if light from a supernova were scattered by dust particles in the intergalactic space, the supernova would look dimmer, and we would be fooled into thinking that it was farther away. These doubts were dispelled a few years later, when Adam Riess of the Space Telescope Science Institute in Baltimore analyzed the most distant supernova known at that time, SN 1997ff. If the dimming were due to obscuration by dust, the effect would only increase with the distance. But this supernova was *brighter*, not dimmer, than it would be in a "coasting" universe that neither accelerates nor decelerates. The explanation was that it exploded at 3 billion years A.B., during the epoch when the vacuum energy was still subdominant and the accelerated expansion had not yet begun.

As the evidence for cosmic acceleration was getting stronger, cosmologists were quick to realize that from certain points of view, the return of the

---

*Mukhanov is the same fellow who first calculated the density perturbations resulting from quantum processes during inflation (see his photo on p. 60).

cosmological constant was not such a bad thing. First, as we discussed in Chapter 9, it provided the missing mass density to make the total density of the universe equal to the critical density. And second, it resolved the nagging cosmic age discrepancy. The age of the universe calculated without a cosmological constant turns out to be smaller than the age of the oldest stars. Now, if the cosmic expansion accelerates, then it was slower in the past, so it took the universe longer to expand to its present size.* The cosmological constant, therefore, makes the universe older, and the age discrepancy is removed.[9]

Thus, only a few years after cosmic acceleration was discovered, it was hard to see how we could ever live without it. The debate now shifted to understanding what it actually meant.

EXPLAINING THE COINCIDENCE

The observed value of the vacuum energy density, about three times the average matter density, was in the ballpark of values predicted three years earlier from the principle of mediocrity. Normally, physicists regard a successful prediction as strong evidence for the theory. But in this case they were not in a hurry to give anthropic arguments any credit. In the years following the discovery, there was a tremendous effort by many physicists to explain the accelerated expansion without relying on the anthropics. The most popular of these attempts was the *quintessence* model, developed by Paul Steinhardt and his collaborators.[10]

The idea of quintessence is that the vacuum energy is not a constant, but is gradually decreasing with the expansion of the universe. It is so small now because the universe is so old. More specifically, quintessence is a scalar field whose energy landscape looks as if it were designed for downhill skiing (Figure 14.3). The field is assumed to start high up the hill in the early universe, but by now it has rolled down to low elevations—which means low energy densities of the vacuum.

The problem with this model is that it does not resolve the coincidence puzzle: why the present energy density of the vacuum happens to be com-

---

*Here, the term "universe" is used in the sense of "visible universe," and "the age of the universe" in the sense of "the time since the big bang in our local region."

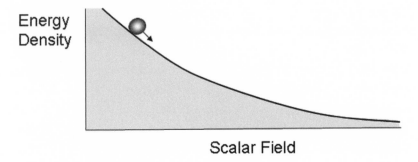

Figure 14.3. Quintessence energy landscape.

parable to the matter density (see Chapter 12). The shape of the energy hill can be adjusted for this to happen, but that would amount to simply fitting the data, instead of explaining it.[11]

On the other hand, the anthropic approach naturally resolves the puzzle. According to the principle of mediocrity, most observers live in regions where the cosmological constant caught up with the density of matter at about the epoch of galaxy formation. The assembly of giant spiral galaxies like ours was completed in the relatively recent cosmological past, at several billion years A.B.[12] Since then, the density of matter has fallen below that of the vacuum, but not by much (by a factor of 3 or so in our region).[13]

Despite numerous attempts, no other plausible explanations for the coincidence have been suggested. Gradually, the collective psyche of the physicists was getting used to the thought that the anthropic picture might be here to stay.

PROS AND CONS

The reluctance of many physicists to embrace the anthropic explanation is easy to understand. The standard of accuracy in physics is very high, you might say unlimited. A striking example of that is the calculation of the magnetic moment of the electron. An electron can be pictured as a tiny magnet. Its magnetic moment, characterizing the strength of the magnet, was first calculated by Paul Dirac in the 1930s. The result agreed very well with experiments, but physicists soon realized that there was a small correction to Dirac's value, due to quantum fluctuations of the vacuum. What followed was a race between particle theorists doing more and more accurate calculations and experimentalists measuring the magnetic moment with higher and higher precision. The most recent measurement result for the correction factor is 1.001159652188, with some uncertainty in the last digit. The theoretical calculation is even more accurate. Remarkably, the agreement between the two is up to the eleventh decimal point. In fact, failure to agree at this level would be a cause for alarm, since any disagreement, even in the eleventh decimal point, would indicate some gap in our understanding of the electron.

Anthropic predictions are not like that. The best we can hope for is to calculate the statistical bell curve. Even if we calculate it very precisely, we will only be able to predict some range of values at a specified confidence level. Further improvements in the calculation will not lead to a dramatic increase in the accuracy of the prediction. If the observed value falls within the predicted range, there will still be a lingering doubt that this happened by sheer dumb luck. If it doesn't, there will be doubt that the theory might still be correct, but we just happened to be among a small percentage of observers at the tails of the bell curve.

It's little wonder that, given a choice, physicists would not give up their old paradigm in favor of anthropic selection. But nature has already made her choice. We only have to find out what it is. If the constants of nature vary from one part of the universe to another, then, whether we like it or not, the best we can do is to make statistical predictions based on the principle of mediocrity.

The observed value of the cosmological constant gives a strong indication that there is indeed a huge multiverse out there. It is within the range of values predicted from anthropic considerations, and there seem to be no credible alternatives. The evidence for the multiverse is, of course, indirect, as it will always be. This is a circumstantial case, where we are not going to hear eyewitness accounts or see the murder weapon. But if, with some luck, we make a few more successful predictions, we may still be able to prove the case beyond a reasonable doubt.

15

A Theory of Everything

What I am really interested in is whether God could have made the world in a different way; that is, whether the necessity of logical simplicity leaves any freedom at all.

—ALBERT EINSTEIN

IN SEARCH OF THE FINAL THEORY

The anthropic picture of the world hinges on the assumption that the constants of nature can vary from one place to another. But can they really? This is a question about the fundamental theory of nature: Will it yield a unique set of constants, or will it allow a wide range of possibilities?

We don't know what the fundamental theory is, and there is no guarantee that it really exists, but the quest for the final, unified theory inspires much of the current research in particle physics. The hope is that beneath the plurality of particles and the differences between the four basic interactions, there is a single mathematical law that governs all elementary phenomena. All particle properties and the laws of gravitation, electromagnetism, and strong and weak interactions would follow from this law, just as all theorems of geometry follow from Euclid's five axioms.

The kind of explanation for the particle properties that physicists hope to find in the final theory is well illustrated by how the chemical properties of the elements were explained in quantum mechanics. Early in the last century, atoms were thought to be the fundamental building blocks of matter. Each type of atom represents a different chemical element, and chemists had accumulated a colossal amount of data on the properties of the elements and their interactions with one another. Ninety-two different elements were known at the time—a bit too many, you might say, for the fundamental building blocks. Thankfully, the work of the Russian chemist Dmitry Mendeleyev in the late nineteenth century revealed some regularity in this mountain of data. Mendeleyev arranged elements in a table in order of increasing atomic weight and noticed that elements with similar chemical properties appeared at regular intervals throughout the table.* Nobody could explain, however, why the elements followed this periodic pattern.

By 1911 it became clear that atoms were not fundamental after all. Ernest Rutherford demonstrated that an atom consisted of a swarm of electrons orbiting a small, heavy nucleus. A quantitative understanding of the atomic structure was achieved in the 1920s, with the development of quantum mechanics. It turned out, roughly, that electron orbits form a series of concentric shells around the nucleus. Each shell can hold only up to a certain number of electrons. So, as we add more electrons, the shells gradually fill up. The chemical properties of an atom are determined mainly by the number of electrons in its outermost shell. As a new shell is filling up, the properties of the elements follow closely those of the preceding shell.† This explains the periodicity of Mendeleyev's table.

For a few brief years it seemed that the fundamental structure of matter was finally understood. Paul Dirac, one of the founders of quantum mechanics, declared in his 1929 paper that "the underlying physical laws necessary for the mathematical theory of a larger part of physics and the whole of chemistry are thus completely known." But then, one by one, new "elementary" particles began to pop up.

To start with, the atomic nuclei turned out to be composite, consisting of protons and neutrons held together by the strong nuclear force. Then the

---

*Another important contribution that Mendeleyev gave to humanity was perfecting the recipe for Russian vodka.

†In other words, any two atoms with a different number of populated shells, but with the same number of electrons in the outer shell, will display similar chemical behavior.

positron was discovered, and following that the muon.* When protons were smashed into one another in particle accelerators, new types of short-lived particles showed up. This did not necessarily mean that protons were made of those particles. If you smash two TV sets together, you can be sure that the things flying out in the debris are the parts the TV sets were originally made of. But in the case of colliding protons, some of the resulting particles were heavier than the protons themselves, with the extra mass coming from the kinetic energy of motion of the protons. So, these collision experiments did not reveal much about the internal structure of the proton, but simply extended the particle zoo. By the end of the 1950s, the number of particles exceeded that of the known chemical elements.† One of the pioneers of particle physics, Enrico Fermi, said that if he could remember the names of all the particles, he would become a botanist.[1]

The breakthrough that introduced order into this unruly crowd of particles was made in the early 1960s, independently, by Murray Gell-Mann of Caltech and Yuval Ne'eman, an Israeli military officer who took leave to complete his Ph.D. in physics. They noticed that all strongly interacting particles fell into a certain symmetric pattern. Gell-Mann and independently George Zweig of CERN (the European Centre for Nuclear Research) later showed that the pattern could be neatly accounted for if all these particles were composed of more fundamental building blocks, which Gell-Mann called *quarks*. This reduced the number of elementary particles, but not by much: quarks come in three "colors" and six "flavors," so there are eighteen different quarks and as many antiquarks. Gell-Mann was awarded the 1969 Nobel Prize for uncovering the symmetry of strongly interacting particles.

In a parallel development, a somewhat similar symmetry was discovered for the particles interacting through weak and electromagnetic forces. The key role in the formulation of this electroweak theory was played by Harvard physicists Sheldon Glashow and Steven Weinberg and the Pakistani physicist Abdus Salam. They shared the 1979 Nobel Prize for this work. Classification of particles according to symmetries played a role analogous to that of the periodic table in chemistry. In addition, three types of

---

*Positrons are antiparticles of the electrons. Muons are unstable particles, very similar to electrons, but 200 times heavier.

†Most of these new particles are unstable and decay into the familiar, stable particles after a brief period of time.

"messenger" particles, which mediate the three basic interactions, were identified: photons for the electromagnetic force, W and Z particles for the weak force, and eight *gluons* for the strong force. All these ingredients provided a basis for the *Standard Model* of particle physics.

The development of the Standard Model was completed in the 1970s. The resulting theory gave a precise mathematical scheme that could be used to determine the outcomes of encounters between any known particles. This theory has been tested in countless accelerator experiments, and as of now it is supported by all the data. The Standard Model also predicted the existence and properties of W and Z particles and of an additional quark—all later discovered. By all these accounts, it is a phenomenally successful theory.

And yet, the Standard Model is obviously too baroque to qualify as the ultimate theory of nature. The model contains more than sixty elementary particles—not a great improvement over the number of elements in Mendeleyev's table. It includes nineteen adjustable parameters, which had to be determined from experiments but are completely arbitrary as far as the theory is concerned. Furthermore, one important interaction—gravity—is left out of the model.[2] The success of the Standard Model tells us that we are on the right track, but its shortcomings indicate that the quest should continue.[3]

THE PROBLEM WITH GRAVITY

The omission of gravity in the Standard Model is not just an oversight. On the face of it, gravity appears to be similar to electromagnetism. Newton's gravitational force, for example, has the same inverse square dependence on the distance as Coulomb's electric force. However, all attempts to develop a quantum theory of gravity along the same lines as the theory of electromagnetism, or other interactions in the Standard Model, encountered formidable problems.

The electric force between two charged particles is due to a constant exchange of photons. The particles are like two basketball players who pass the ball back and forth to one another as they run along the court. Similarly, the gravitational interaction can be pictured as an exchange of gravitational field quanta, called *gravitons*. And indeed, this description works rather well, as long as the interacting particles are far apart. In this case, the gravitational force is weak and the spacetime is nearly flat. (Remember, gravity is related to the curvature of spacetime.) The gravitons can be pictured as little humps bouncing between the particles in this flat background.

At very small distances, however, the situation is completely different. As we discussed in Chapter 12, quantum fluctuations at short distance scales give the spacetime geometry a foamlike structure (see Figure 12.1). We have no idea how to describe the motion and interaction of particles in such a chaotic environment. The picture of particles moving through a smooth spacetime and shooting gravitons at one another clearly does not apply in this regime.

Effects of quantum gravity become important only at distances below the Planck length—an unimaginably tiny length, 10^{25} times smaller than the size of an atom. To probe such distances, particles have to be smashed at tremendous energies, far beyond the capabilities of the most powerful accelerators. On much larger distance scales, which are accessible to observation, quantum fluctuations of spacetime geometry average out and quantum gravity can be safely ignored. But the conflict between Einstein's general relativity and quantum mechanics cannot be ignored in our search for the ultimate laws of nature. Both gravity and quantum phenomena have to be accounted for in the final theory. Thus, leaving gravity out is not an option.

THE HARMONY OF STRINGS

Most physicists now place their hopes on a radically new approach to quantum gravity—the theory of strings. This theory provides a unified description for all particles and all their interactions. It is the most promising candidate we have ever had for the fundamental theory of nature.

According to string theory, particles like electrons or quarks, which seem to be pointlike and were thought to be elementary, are in fact tiny vibrating loops of string. The string is infinitely thin, and the length of the little loops is comparable to the Planck length. The particles appear to be structureless points because the Planck length is so tiny.

The string in little loops is highly taut, and this tension causes the loops to vibrate, in a way similar to the vibrating strings in a violin or piano. Different vibration patterns on a straight string are illustrated in Figure 15.1. In these patterns, which correspond to different musical notes, the string has a wavy shape, with several complete half-waves fitting along its length. The larger the number of half-waves, the higher the note. Vibration patterns of loops in string theory are similar (see Figure 15.2), but now different patterns correspond to different types of particles, rather than different notes.

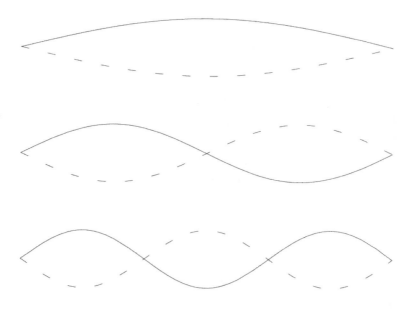

Figure 15.1. Vibration patterns of a straight string.

Figure 15.2. A schematic representation of vibration patterns of a string loop.

The properties of a particle—for example, its mass, electric charge, and charges with respect to weak and strong interactions—are all determined by the exact vibrational state of the string loop. Instead of introducing an independent new entity for each type of particle, we now have a single entity—the string—of which all particles are made.

The messenger particles—photons, gluons, W, and Z—are also vibrating little loops, and particle interactions can be pictured as string loops splitting and joining. What is most remarkable is that the spectrum of string states necessarily includes the graviton—the particle mediating gravitational interactions. The problem of unifying gravity with other interactions does not exist in string theory: on the contrary, the theory cannot be constructed without gravity.

The conflict between gravity and quantum mechanics has also disappeared. As we have just discussed, the problem was due to quantum fluctuations of the spacetime geometry. If particles are mathematical points, the fluctuations go wild in the vicinity of the particles and the smooth spacetime continuum turns into a violent spacetime foam. In string theory, the little loops of string have a finite size, which is set by the Planck length. This is precisely the distance scale below which quantum fluctuations get out of control. The loops are immune to such sub-Planckian fluctuations: the spacetime foam is tamed just when it is about to start causing trouble. Thus, for the first time we have a consistent quantum theory of gravity.

The idea that particles may secretly be strings was suggested in 1970 by Yoichiro Nambu of the University of Chicago, Holger Nielsen of the Niels Bohr Institute, and Leonard Susskind of Yeshiva University. String theory was first meant to be a theory of strong interactions, but it was soon found that it predicted the existence of a massless boson, which had no counterpart among the strongly interacting particles. The key realization that the massless boson had all the properties of the graviton was made in 1974 by John Schwarz of Caltech and Joel Scherk of the École Normale Supérieure. It took another ten years for Schwarz, working in collaboration with Michael Green of Queen Mary College in London, to resolve some subtle mathematical issues and show that the theory was indeed consistent.

String theory has no arbitrary constants, so it does not allow any tinkering or adjustments. All we can do is to uncover its mathematical framework and see whether or not it corresponds to the real world. Unfortunately, the mathematics of the theory is incredibly complex. Now, after twenty years of assault by hundreds of talented physicists and mathematicians, it is still far from being fully understood. At the same time, this research revealed a mathematical structure of amazing richness and beauty. This, more than anything else, suggests to the physicists that they are probably on the right track.[4]

THE LANDSCAPE

As I have just mentioned, string theory has no adjustable parameters. This is not an exaggeration: I mean *none* at all. Even the number of the dimensions of space is rigidly fixed by the theory. The problem is that it gives the wrong answer: it requires that space should have nine dimensions instead of three.

This sounds very embarrassing: Why should we even consider a theory that is in such blatant conflict with reality? The conflict, however, can be avoided if the extra six dimensions are curled up or, as physicists say, *compactified*. A soda straw is a simple example of compactification: it has one large dimension along the straw and another curled up in a small circle. When viewed from a distance, the straw looks like a one-dimensional line, but close by we can see that its surface is in fact a two-dimensional cylinder (see Figure 15.3). Quite similarly, the compact extra dimensions will not be visible if they are sufficiently small. In string theory, they are not expected to be much larger than the Planck length.[5]

The main problem with extra dimensions is that it is not clear exactly how they are to be compactified. If there were a single extra dimension, there would be only one way to compactify it: to curl it up into a circle. Two extra dimensions can be compactified as a sphere, as a doughnut, or as a more

Figure 15.3. A soda straw has a two-dimensional cylindrical surface. It has a large dimension along the straw and a small dimension curled up in a circle.

complicated surface with a large number of "handles" (Figure 15.4). As you go to higher dimensions, the number of possibilities multiplies. The vibrational states of the strings depend on the size and shape of extra dimensions, so each new compactification corresponds to a new vacuum with different types of particles, having different masses and different interactions.

The hope of string theorists was that in the end the theory would yield a unique compactification describing our world and we would finally have an explanation for the observed values of all the particle physics parameters.[6] On the wave of excitement that followed some mathematical breakthroughs in the 1980s, it seemed that this goal might be just around the corner, and string theory was heralded as the future "theory of everything"— a tall order for a theory that was yet to make its first observational prediction! But gradually, a very different picture was beginning to emerge: the theory appeared to allow thousands of different compactifications.

If this was not bad enough, things got considerably worse in the mid-1990s as a result of some unexpected new developments. As the mathematics of string theory was better understood, it became clear that in addition to one-dimensional strings, the theory must include two-dimensional membranes, as well as their higher-dimensional analogues. All these new arrivals

Figure 15.4. Different ways to compactify two extra dimensions. The large, noncompact dimensions are not shown.

are collectively called *branes*.* Vibrating little branes would look like particles, but they are too massive to be produced in particle accelerators.[7]

The branes have one unpleasant side effect: they dramatically increase the number of ways in which new vacua can be constructed. A brane can be wrapped, like a rubber band, around some of the compact dimensions. Every new stable brane configuration gives a new vacuum. You can wrap one, two, or more branes on each of the handles of the compact space, and with a large number of handles, the number of possibilities is enormous. The equations of the theory have no adjustable constants, but their solutions, describing different vacuum states, are characterized by several hundred parameters—the sizes of compact dimensions, the locations of the branes, and so on.

If we had just one parameter, it would be very similar to a scalar field in the usual particle theory. As we discussed in earlier chapters, it would then behave as a little ball in the energy landscape and would roll to the nearest minimum of the energy density. With two parameters, the landscape would be two-dimensional, as illustrated in Figure 15.5. It would have maxima (peaks) and minima (valleys), with minima representing the vacuum states. The altitude at each minimum gives the corresponding vacuum energy density (the cosmological constant).

The actual energy landscape of string theory is much more complicated, since it includes many more parameters. This landscape cannot be drawn on a sheet of paper: to account for all the parameters, we would need a space of several hundred dimensions. But the landscape can still be mathematically analyzed. A rough estimate indicates that it contains about 10^{500} (google to the fifth power!) different vacua. Some of these vacua are similar to ours; others have very different values for the constants of nature. Still others differ more drastically and have totally different kinds of particles and interactions, or more than three large dimensions.

As the outlines of the landscape were emerging, the hope of deriving a unique vacuum from string theory was rapidly slipping away. But string theorists were in denial and not ready to accept defeat.

---

*The theory also includes a host of other entities (e.g., *fluxes*, which are similar to magnetic fields), but I will omit them in this discussion.

Figure 15.5. Energy landscape in two dimensions. Each dimension (not to be confused with the dimensions of ordinary space) represents one of the parameters characterizing string theory vacua.

THE BUBBLING UNIVERSE

The first physicists to break from the pack were Raphael Bousso, now at the University of California at Berkeley, and Joseph Polchinski of the Kavli Institute for Theoretical Physics at Santa Barbara. Remember Polchinski? He is the string theorist who could not stand the anthropic principle and pledged to quit physics if the cosmological constant was discovered.* Luckily, he changed his mind—both about quitting physics and about the anthropic principle.

Bousso and Polchinski combined the picture of the string theory landscape with the ideas of inflationary cosmology and argued that regions of all possible vacua will be created in the course of eternal inflation. The highest-energy vacuum will inflate the fastest. Bubbles of lower-energy vacua will nucleate and expand in this inflating background (as in Guth's

---

*Polchinski is largely responsible for the realization that string theory must include branes of different dimensions.

Figure 15.6. Bubbles filled with lower-energy vacua nucleate in the inflating high-energy background, and still-lower energy bubbles nucleate inside them.

original inflationary scenario, discussed in Chapters 5 and 6). The interiors of the bubbles will inflate at a smaller rate, and bubbles of still-lower energy will pop out inside them (see Figure 15.6).* As a result, the entire string theory landscape will be explored—countless bubbles will be formed, filled with every possible kind of vacuum.[8]

We live in one of the bubbles, but the theory does not tell us which one. Only a tiny fraction of the bubbles are hospitable to life, and we must find ourselves in one of these rare bubbles. Much to the dismay of many string theorists, this is precisely the kind of picture that is assumed in anthropic arguments. If string theory is indeed the ultimate theory of reality, then it appears that the anthropic worldview is inevitable.

It needs to be said that the landscape of string theory is far from being fully mapped. In order to yield a realistic cosmology, some of the slopes have to be very gentle, allowing for slow-roll inflation. Recent work indi-

---

*Bubbles of *higher* energy density can also be formed, though with a much smaller probability.

cates that there are indeed such regions in the landscape. We should also search for even gentler slopes, required by Linde's scalar field model of a variable cosmological "constant" (discussed in Chapter 13). None have been found so far. But Bousso and Polchinski suggest that googles of vacua in the landscape provide a suitable alternative.

Instead of a continuum of vacuum energy densities in Linde's model, the landscape gives a discrete set of values. Normally, this would be a problem, because only a tiny fraction of these values (about 1 in 10^{120}) fall in the small anthropically allowed range. If we had less than 10^{120} vacua, this range would most probably be empty. But with 10^{500} vacua in the landscape, the set of values is so dense that it is almost continuous, and we expect that googles of vacua will have the cosmological constant in the anthropically allowed interval. The principle of mediocrity can then be applied in the same manner as before, and the successful prediction of the observed cosmological constant is unaffected.

A PROGRAM FOR THE TWENTY-FIRST CENTURY

The paper by Bousso and Polchinski, which appeared in 2000, did make a stir, but the landslide began three years later, when they were joined by one of the inventors of string theory, Leonard Susskind, then at Yeshiva University. Susskind is a fiercely independent thinker and is also a man of great charm and charisma. His power of persuasion is phenomenal; this is the man you want to have on your side.

Susskind was still unconvinced when Bousso and Polchinski's paper first came out. He felt that the existence of a multitude of vacua assumed in the paper relied more on conjecture than on mathematical fact. But the developments of the following few years showed that the conjectures were basically sound, and in 2003 Susskind came out in full force promoting what he called "the anthropic landscape of string theory." He argued that the diversity of vacua in string theory provided, for the first time, a solid scientific basis for anthropic arguments. String theorists, he said, should therefore embrace the anthropic principle, instead of fighting against it.

In less than a year, everybody was talking about "the landscape." The number of papers discussing multiple vacua and other anthropic-related issues grew from four in 2002 to thirty-two in 2004. Of course, not everybody was pleased with this turn of events. "I hate this recent landscape idea," says Paul Steinhardt, "and I am hopeful it will go away."[9] David Gross, the 2004 Nobel Prize winner, who regards the use of the anthropic principle as giving up the ideal of uniqueness, paraphrased Winston Churchill, saying "Never, never, never, never give up!" When I talked to him at a meeting in Cleveland, he complained that the anthropic principle is like a virus. Once you get it, you are lost to the community. "Ed Witten* dislikes this idea intensely," says Susskind, describing the situation, "but I'm told he's very nervous that it might be right. He's not happy about it, but I think he knows that things are going in that direction."[10]

If the landscape ideas are correct, explaining the observed constants of nature is not going to be easy. First, we will need to map the landscape. What kinds of vacua are there, and how many of each kind? We cannot realistically hope to obtain a detailed characterization of all 10^{500} vacua, so some kind of statistical description will be necessary. We will also need to estimate the probabilities for bubbles of one vacuum to form amidst another vacuum. Then we will have all the ingredients to develop a model of an eternally inflating universe with bubbles inside bubbles inside bubbles, as illustrated in Figure 15.6. Once we have this model, the principle of mediocrity can be used to determine the probability for us to live in one of the vacua or another.

We are now making our first, tentative steps in this program, and for-midable challenges lie ahead. "But," writes Leonard Susskind, "I would bet that at the turn of the 22nd century, philosophers and physicists will look nostalgically at the present and recall a golden age in which the narrow provincial 20th century concept of the universe gave way to a bigger better megaverse, populating a landscape of mind-boggling proportions."[11]

---

*Edward Witten, one of the leading string theorists, was awarded the 1990 Fields Medal—the mathematics equivalent of the Nobel Prize.

PART IV

. . .

BEFORE THE BEGINNING

16

Did the Universe Have a Beginning?

*Whence all creation had its origin, . . . he, who surveys it all
from highest heaven, he knows or maybe even he does not know.*
—Rig-Veda

A PROBLEM WITH THE COSMIC EGG

Ancient creation myths display wonderful ingenuity, but at the most
fundamental level they have to choose one of two basic options: either
the universe was created a finite time ago, or it has existed forever.[1]

Here is one of several scenarios offered in the sacred Hindu scripture,
the Upanishads:

> In the beginning this [world] was nonexistent. It became existent. It
> turned into an egg. The egg lay for the period of a year. Then it broke
> open . . . And what was born of it was yonder Aditya, the Sun. When it
> was born shouts of "Hurrah" arose, together with all beings and all ob-
> jects of desire.

This idea looks simple enough, but unfortunately it has a serious flaw, which
it shares with every other story of creation. The ancients were well aware of
the problem; the Jain poet Jinasena wrote in the ninth century:

The doctrine that the world was created is ill-advised, and should be rejected.

If God created the world, where was he before creation? . . .

How could God have made the world without any raw material? If you say he made this first, and then the world, you are faced with an endless regression . . .

Thus the doctrine that the world was created by God makes no sense at all . . .

Know that the world is uncreated, as time itself is, without beginning and end . . . Uncreated and indestructible, it endures under the compulsion of its own nature.[2]

This critique applies equally well to every scenario of the cosmic origin—be it a creation by God, as in the story of the cosmic egg, or a "natural" creation, such as the big bang model of modern cosmology.[3]

According to the big bang theory, all the matter that we see around us came out of a hot cosmic fireball some 14 billion years ago. But where did the fireball come from? The theory of inflation has shown that an expanding fireball could arise out of a tiny false-vacuum nugget. But the question still remains: Where did that initial nugget originate? What happened before inflation?

For the most part, cosmologists were in no hurry to tackle this thorny issue. In fact, it appeared that a satisfactory answer could never be given. Whatever the answer is, one can always ask "And what happened before that?" This is the "endless regression" that Jinasena is referring to. However, in the 1980s, when the eternal inflation scenario was developed, it appeared to offer an attractive alternative.

An eternally inflating universe consists of an expanding "sea" of false vacuum, which is constantly spawning "island universes" like ours. Inflation is thus a never-ending process. It has ended in our own island universe, but will continue indefinitely in other remote regions. But if inflation is going to continue forever into the future, then perhaps it might have had no beginning in the past. We would then have an eternally inflating universe without a beginning and without an end; that would eliminate the perplexing problems associated with the cosmic origin. This picture is reminiscent of the steady-state cosmology of the 1940s and '50s. Some people found it very appealing.

A CYCLIC UNIVERSE

Apart from a steady state, there is another way for the universe to be eternal. And again, the Hindus figured this out a long time ago. The endless cycle of creation and destruction is symbolized by the dance of the god Shiva. "He rises from His rapture and, dancing, sends through inert matter pulsing waves of awakening sound." The universe comes to life, but then "[i]n the fullness of time, still dancing, He destroys all forms and names by fire and gives new rest."[4]

A parallel idea in scientific cosmology is that of an oscillating universe, which goes through a cycle of expansion and contraction. It was briefly popular in the 1930s, but then fell out of favor, because of apparent conflict with the second law of thermodynamics.

The second law requires that entropy, which is a measure of disorder, should grow in each cycle of cosmic evolution. If the universe had already gone through an infinite number of cycles, it would have reached the maximum-entropy state of thermal equilibrium. We certainly do not find ourselves in such a state. This is the "heat death" problem that I mentioned earlier.

The idea of an oscillating universe was abandoned for more than half a century, but in 2002 it was revived in a new guise by Paul Steinhardt and Neil Turok of Cambridge University. As in earlier models, they suggested that the history of the universe consists of an endlessly repeating cycle of expansion and contraction. Each cycle starts with a hot expanding fireball. It expands and cools down, galaxies form, and the vacuum energy comes to dominate the universe soon thereafter. At this point the universe starts expanding exponentially, with its size doubling every 10 billion years or so. After trillions of years of this super-slow inflation, the universe becomes very homogeneous, isotropic, and flat. Eventually the expansion slows down and then turns into contraction. The universe recollapses and immediately bounces back to start a new cycle. Part of the energy generated in the collapse goes to create a hot fireball of matter.[5]

Steinhardt and Turok argue that the problem of the beginning does not arise in their scenario. The universe has always been going through the same cycle, so there was no beginning. The problem of the heat death is also avoided, because the amount of expansion in a cycle is greater than the amount of contraction, so the volume of the universe is increased after each cycle. The entropy of our observable region is now the same as the entropy of a similar region in the preceding cycle, but the entropy of the entire universe has increased, simply because the volume of the universe is now greater. As time goes on, both the entropy and the total volume grow without bound. The state of maximum entropy is never reached, because there is no maximum entropy.

Thus, it appears that we have two possible models for an eternal universe without a beginning: one is based on eternal inflation and the other on cyclic evolution. However, it turns out that neither possibility can yield a complete description of the universe.

DE SITTER SPACE

When a physicist wants to understand some phenomenon, the first thing she does is to maximally simplify it, stripping it down to the bare essentials. In the case of eternal inflation, we can strip away island universes, keeping only the inflating sea. In addition, we can assume that the universe is homogeneous and isotropic, as in Friedmann's models. With these simplifications, Einstein's equations for the inflating universe can be easily solved.

The solution has the geometry of a three-dimensional sphere, which contracts from a very large radius in the remote past. The contraction is slowed down by the repulsive gravity of the false vacuum, until the sphere stops for a moment and then starts to re-expand. The force of gravity now works in the direction of motion, so the sphere expands with acceleration. Its radius grows exponentially, with a doubling time determined by the energy density of the false vacuum.*

The solution I have just described has been known since the early days of general relativity; it is called *de Sitter spacetime*—after the Dutch as-

---

*The minimum radius of the de Sitter sphere is roughly equal to the distance traveled by light during one doubling time of inflation.

tronomer Willem de Sitter, who discovered it in 1917. This spacetime is illustrated in Figure 16.1. Inflation begins in de Sitter spacetime only after the spherical universe has reached its minimum radius. But once started, the exponential expansion continues forever, so inflation is eternal to the future.

If we were to allow the formation of island universes, they would collide and merge in the contracting part of spacetime. The islands would then quickly fill the entire space, the false vacuum would be completely eliminated, and the universe would continue collapsing to a big crunch. Thus, inflation cannot be extended into the infinite past. It must have had some sort of beginning.

We should keep in mind, however, that this conclusion is based on the maximally simplified model of inflation, which assumes a homogeneous and isotropic universe. In reality, the universe may well be very irregular—inhomogeneous and anisotropic—on scales much greater than the present horizon. Could it be, then, that the contracting phase of de Sitter spacetime is an artifact of the simplifying assumptions that we have made? Is it possible to avoid the beginning in a more general spacetime?

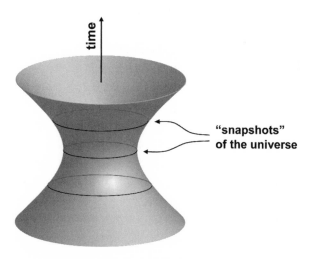

Figure 16.1. De Sitter spacetime, with two of the three spatial dimensions suppressed. Horizontal slices of the spacetime give "snapshots" of the universe at different moments of time. In a four-dimensional spacetime the slices would be three-dimensional spherical spaces.

BEYOND UNREASONABLE DOUBT

These doubts were put to rest only recently, in a paper I wrote in collaboration with Arvind Borde of Long Island University and Alan Guth. The theorem we proved in that paper is amazingly simple. Its proof does not go beyond high school mathematics, but its implications for the beginning of the universe are very profound.

Borde, Guth, and I studied what an expanding universe looks like from the point of view of different observers. We considered imaginary observers moving through the universe under the action of gravity and inertia and recording what they see. If the universe had no beginning, then the histories of all such observers should extend into the infinite past. We showed that this assumption leads to a contradiction.

To have a specific picture in mind, suppose there is an observer in every galaxy of our local region. Since the universe is expanding, each of these observers will see the others moving away. Galaxies may not exist in some regions of space and time, but we can still imagine the entire universe "sprinkled" with observers in such a way that all of them are moving away from one another.* To give these observers some name, we shall call them "spectators."

Let us now introduce another observer who is moving relative to the spectators. We shall call him the space traveler. He is moving by inertia, with the engines of his spaceship turned off, and has been doing so for all eternity. As he passes the spectators, they register his velocity.

Since the spectators are flying apart, the space traveler's velocity relative to each successive spectator will be smaller than his velocity relative to the preceding one. Suppose, for example, the space traveler has just zoomed by the Earth at the speed of 100,000 kilometers per second and is now headed toward a distant galaxy, about a billion light-years away. That galaxy is moving away from us at 20,000 kilometers per second, so when the space traveler catches up with it, the observers there will see him moving at 80,000 kilometers per second.

---

*The existence of such a class of observers can be taken as a definition of an expanding universe.

If the velocity of the space traveler relative to the spectators gets smaller and smaller into the future, it follows that his velocity should get larger and larger as we follow his history into the past. In the limit, his velocity should get arbitrarily close to the speed of light.

The key insight of my paper with Borde and Guth is that as we go into the past and approach past infinity by the clocks of the spectators, the time elapsed by the clock of the space traveler is still *finite*. The reason is that according to Einstein's theory of relativity, a moving clock ticks slower, and the closer you get to the speed of light, the more slowly it ticks. As we go backward in time, the speed of the space traveler approaches the speed of light and his clock essentially comes to a halt. This is from the spectator's point of view. But the space traveler himself does not notice anything unusual. For him, what spectators perceive as a frozen moment, stretched into eternity, is a moment like any other, which has to be preceded by earlier moments. Like the histories of the spectators, the space traveler's history should extend into the infinite past.

The fact that the time elapsed by the space traveler's clock is finite indicates that we do not have his full history. This means that some part of the past history of the universe is missing; it is not included in the model. Thus, the assumption that the entire spacetime can be covered by an expanding "dust" of observers has led to a contradiction, and therefore it cannot be true.[6]

A remarkable thing about this theorem is its sweeping generality. We made no assumptions about the material content of the universe. We did not even assume that gravity is described by Einstein's equations. So, if Einstein's gravity requires some modification, our conclusion will still hold. The only assumption that we made was that the expansion rate of the universe never gets below some nonzero value, no matter how small.[7] This assumption should certainly be satisfied in the inflating false vacuum. The conclusion is that past-eternal inflation without a beginning is impossible.

What about a cyclic universe? It has alternating periods of expansion and contraction. Can this help the universe to escape from the clutches of the theorem? The answer turns out to be no. An essential feature of the cyclic scenario, which allows it to avoid the heat-death problem, is that the volume of the universe increases in every cycle, so on average the universe is expanding. In my paper with Borde and Guth, we show that as a result of this expansion, the space traveler's velocity increases on average as we go back in time and still approaches the speed of light in the limit. Hence, the same conclusions apply.[8]

It is said that an argument is what convinces reasonable men and a proof is what it takes to convince even an unreasonable man. With the proof now in place, cosmologists can no longer hide behind the possibility of a past-eternal universe. There is no escape: they have to face the problem of a cosmic beginning.

•

Working with Alan Guth on a paper was a memorable experience. The idea of the proof emerged in e-mail exchanges between me, Alan, and Arvind, and the details were nailed down in two hours at the blackboard, when the three of us met in my Tufts office in August 2001. In about a month, we wrote a paper and submitted it to *Physical Review Letters*. I was amazed. What happened to Alan and his legendary procrastination? But I was not to be disappointed. In a few months, the editor sent us the report of a referee, asking us to clarify some points in the proof. And that is when the good old Alan returned in full glory. His e-mails started arriving at longer and longer intervals, with headings like "swamped at the moment" and "nothing done yet." When he did find some time to work on the paper, he seems to have spent a fair fraction of it on issues such as whether we should thank "an anonymous referee" or "the anonymous referee" for his or her comments. He gave a detailed discussion of pros and cons for either version. Alan might have suspected that his editing of the paper was taking a bit too long, and at some point he wrote "I need to thank you guys for not shooting me." In fairness, I should add that he spent some time on more substantive issues as well and that the drawn-out process of revising the paper resulted in its substantial improvement. The paper was finally published in April 2003.[9]

A PROOF OF GOD?

Theologians have often welcomed any evidence for the beginning of the universe, regarding it as evidence for the existence of God. In the 1950s the accumulating evidence for the big bang inspired enthusiasm in theological circles and among some religiously inclined scientists. "As to the first cause of the universe," wrote the British physicist Edward Milne, "in the context of expansion, this is left for the reader to insert, but our picture is incomplete without Him."[10] The big bang theory even received an official endorsement from the Roman Catholic Church. In his 1951 address to the Pontifical Academy of Sciences, Pope Pius XII said that it "has confirmed . . . the well-founded deduction as to the epoch when the cosmos came forth from the hands of the Creator. Hence, Creation took place. Therefore there is a Creator. Therefore God exists!"[11]

For the same reasons that made the pope so exuberant, the natural instinct of most scientists has been to reject the idea of a cosmic beginning. "To deny the infinite duration of time," asserted the Nobel Prize–winning German chemist Walter Nernst, "would be to betray the very foundations of science."[12] The beginning of the universe looked too much like a divine intervention; there seemed to be no possibility to describe it scientifically. This was one thing scientists and theologians seemed to agree upon.

So, what do we make of a proof that the beginning is unavoidable? Is it a proof of the existence of God? This view would be far too simplistic. Anyone who attempts to understand the origin of the universe should be prepared to address its logical paradoxes. In this regard, the theorem that I proved with my colleagues does not give much of an advantage to the theologian over the scientist. As evidenced by Jinasena's remarks earlier in this chapter, religion is not immune to the paradoxes of Creation.

Also, the scientists might have been too rash to admit that the cosmic beginning cannot be described in purely scientific terms. True, it is hard to see how this can be done. But things that seem to be impossible often reflect only the limitations of our imagination.

17

Creation of Universes from Nothing

Nothing can be created from nothing.

—Lucretius

INFLATION AT THE END OF THE TUNNEL

Back in 1982, inflation was still a very new field, full of unexplored ideas and challenging problems—a gold mine for an aspiring young cosmologist. The most intriguing of these problems, and perhaps the least relevant for the present state of the universe, was the question of how inflation could have started. An inflating universe quickly "forgets" its initial conditions, so the state at the onset of inflation has little effect on what happens afterward. Thus, if you want to find ways of testing inflation observationally, you should not waste your time worrying about how it began. But the puzzle of the beginning was still there and could not be avoided. It drew me like a magnet.

At first sight, the problem looked relatively simple. We know that a small region of space filled with false vacuum is enough to drive inflation. So, all I had to figure out was how such a region could have arisen from some earlier state of the universe.

The prevailing view at the time was based on the Friedmann model, where the universe expanded from a singular state of infinite curvature and infinite matter density. Assuming that the universe is filled with a high-energy false vacuum, any matter that was initially present is diluted, and the vacuum energy eventually dominates. At that point, the repulsive gravity of the vacuum takes over, and inflation begins.

This would be fine, except, Why was the universe expanding to begin with? One of the achievements of inflation was to explain the expansion of the universe. Yet, it looked as if we needed to have expansion before inflation even started. The attractive gravity of matter is initially much stronger than the gravitational repulsion of the vacuum, so if we don't postulate a strong initial blast of expansion, the universe would simply collapse and inflation would never begin.

I pondered this argument for a while, but the logic was very simple and there seemed to be no escape. Then, suddenly, I realized that instead of collapsing, the universe could do something much more interesting and dramatic . . .

Suppose we have a closed spherical universe, filled with a false vacuum and containing a certain amount of ordinary matter. Suppose also that this universe is momentarily at rest, neither expanding nor contracting. Its future will depend on its radius. If the radius is small, the matter is compressed to a high density and the universe will collapse to a point. If the radius is large, the vacuum energy dominates and the universe will inflate. Small and large radii are separated by an energy barrier, which cannot be crossed unless the universe is given a large expansion velocity.

What I suddenly realized was that the collapse of a small universe was inevitable only in classical physics. In quantum theory, the universe could *tunnel* through the energy barrier and emerge on the other side—like a nuclear particle in Gamow's theory of radioactive decay.

This looked like a neat solution to the problem. The universe starts out extremely small and is most likely to collapse to a singularity. But there is a small chance that instead of collapsing, it will tunnel through the barrier to a bigger radius and start inflating (see Figure 17.1). So, in the grander scheme of things, there will be loads of failed universes that will exist only for a fleeting moment, but there will also be some that will make it big.

I felt that I was making progress, so I pressed on. Is there any bound to how small the initial universe could be? What happens if we allow it to get smaller and smaller? To my surprise, I found that the tunneling probability

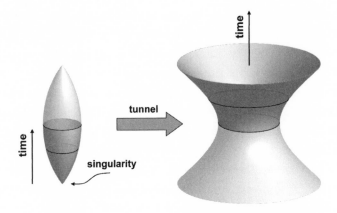

Figure 17.1. On the left, a spacetime diagram of a closed Friedmann universe expanding from a singularity, reaching a maximum radius and recollapsing. Time grows in the vertical direction, and horizontal circles give snapshots of the universe at different moments of time. On the right, a universe dominated by vacuum energy, which contracts and re-expands (de Sitter spacetime). Instead of recollapsing, the universe on the left can tunnel through the energy barrier to a larger radius and start expanding. The spacetime history of the universe will then include only the shaded parts of the two spacetimes.

did not vanish as the initial size approached zero. I also noticed that my calculations were greatly simplified when I allowed the initial radius of the universe to vanish. This was really crazy: what I had was a mathematical description of a universe tunneling from a zero size—from nothing!—to a finite radius and beginning to inflate. It looked as though there was no need for the initial universe!

TUNNELING FROM NOTHING

The concept of a universe materializing out of nothing boggles the mind. What exactly is meant by "nothing"? If this "nothing" could tunnel into something, what could have caused the primary tunneling event? And what about energy conservation? But as I kept thinking about it, the idea appeared to make more and more sense.

The initial state prior to the tunneling is a universe of vanishing radius, that is, no universe at all. There is no matter and no space in this very peculiar state. Also, there is no time. Time has meaning only if something is happening in the universe. We measure time using periodic processes, like the rotation of the Earth about its axis, or its motion around the Sun. In the absence of space and matter, time is impossible to define.

And yet, the state of "nothing" cannot be identified with *absolute* nothingness. The tunneling is described by the laws of quantum mechanics, and thus "nothing" should be subjected to these laws. The laws of physics must have existed, even though there was no universe. I will have more to say about this in Chapter 19.

As a result of the tunneling event, a finite-sized universe, filled with a false vacuum, pops out of nowhere ("nucleates") and immediately starts to inflate. The radius of the newborn universe is determined by the vacuum energy density: the higher the density, the smaller the radius. For a grand-unified vacuum, it is one hundred-trillionth of a centimeter. Because of inflation, this tiny universe grows at a staggering rate, and in a small fraction of a second it becomes much greater than the size of our observable region.

If there was nothing before the universe popped out, then what could have caused the tunneling? Remarkably, the answer is that no cause is required. In classical physics, causality dictates what happens from one moment to the next, but in quantum mechanics the behavior of physical objects is inherently unpredictable and some quantum processes have no cause at all. Take, for example, a radioactive atom. It has some probability of decaying, which is the same from this minute to the next. Eventually, it will decay, but there will be nothing that causes it to decay at that particular moment. Nucleation of the universe is also a quantum process and does not require a cause.

Most of our concepts are rooted in space and time, and it is not easy to create a mental picture of a universe popping out of nothing. You cannot imagine that you are sitting in "nothing" and waiting for a universe to materialize—because there is no space to sit in and there is no time.

In some recently proposed models based on string theory, our space is a three-dimensional membrane (brane) floating in a higher-dimensional space. In such models, we can imagine a higher-dimensional observer watching small bubble universes—braneworlds—pop out here and there, like bubbles of vapor in a boiling pot of water. We live on one of the bubbles, which is an expanding three-dimensional spherical brane. For us, this brane is the only space there is. We cannot get out of it and are unaware of the extra dimensions. As we follow the history of our bubble universe back in time, we come to the moment of nucleation. Beyond that, our space and time disappear.

From this picture, there is only a small step to the one that I originally proposed. Simply remove the higher-dimensional space. From our internal point of view, nothing will change. We live in a closed, three-dimensional space, but this space is not floating anywhere. As we go back in time, we discover that our universe had a beginning. There is no spacetime beyond that.

An elegant mathematical description of quantum tunneling can be obtained using the so-called *Euclidean time*. This is not the kind of time you measure with your watch. It is expressed using imaginary numbers, like the square root of −1, and is introduced only for computational convenience. Making the time Euclidean has a peculiar effect on the character of spacetime: the distinction between time and the three spatial dimensions completely disappears, so instead of spacetime we have a four-dimensional space. If we could live in Euclidean time, we would measure it with a ruler, just as we measure length. Although it may appear rather odd, the Euclidean-time description is very useful: it provides a convenient way to determine the tunneling probability and the initial state of the universe as it emerges into existence.

The birth of the universe can be graphically represented by the spacetime diagram in Figure 17.2. The dark hemisphere at the bottom corresponds to quantum tunneling (time is Euclidean in this part of the spacetime). The light surface above it is the spacetime of the inflating universe. The boundary between the two spacetime regions is the universe at the moment of nucleation.

A remarkable feature of this spacetime is that it has no singularities. A Friedmann spacetime has a singular point of infinite curvature at the beginning, where the mathematics of Einstein's equations breaks down. It is represented by the sharp point (labeled "singularity") at the bottom on the left-hand side of Figure 17.1. In contrast, the Euclidean spherical region has no such points; it has the same finite curvature everywhere. This was the first mathematically consistent description of how the universe could be born. The spacetime diagram of Figure 17.2, which looks a bit like a badminton shuttlecock is now on the logo of the Tufts Institute of Cosmology.

I wrote all this up in a short paper entitled "Creation of Universes from

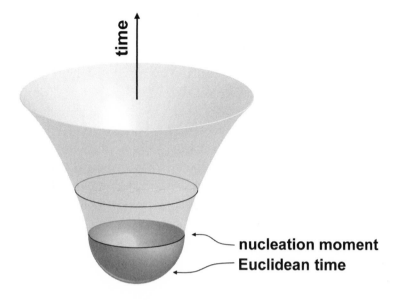

Figure 17.2. A spacetime diagram of the universe tunneling from nothing.

Nothing."[1] Before submitting it to a journal, I made a day trip to Princeton University, to discuss these ideas with Malcolm Perry, a well-known expert on the quantum theory of gravitation. After an hour at the blackboard, Malcolm said, "Well, maybe this is not so crazy . . . How come I have not thought of it myself?" What better compliment can you get from a fellow physicist!

THE UNIVERSE AS A QUANTUM FLUCTUATION

My model of the universe tunneling out of nothing did not appear from nothing—I had some predecessors. The first suggestion of this sort came from Edward Tryon of Hunter College, City University of New York. He proposed the idea that the universe was created out of vacuum as a result of a quantum fluctuation.

The thought first occurred to him in 1970, during a physics seminar. Tryon says that it struck him like a flash of light, as if some profound truth had suddenly been revealed to him. When the speaker paused to collect his thoughts, Tryon blurted out, "Maybe the universe is a vacuum fluctuation!" The room roared with laughter.[2]

As we discussed earlier, the vacuum is anything but dull or static; it is a site of frantic activity. Electric, magnetic, and other fields are constantly fluctuating on subatomic scales because of unpredictable quantum jerks. The spacetime geometry is also fluctuating, resulting in a frenzy of space-time foam at the Planck distance scale. In addition, the space is full of *virtual* particles, which spontaneously pop out here and there and instantly disappear. The virtual particles are very short-lived, because they live on borrowed energy. The energy loan needs to be paid off, and according to Heisenberg's uncertainty principle, the larger the energy borrowed from the vacuum, the faster it has to be repaid. Virtual electrons and positrons typically disappear in about one-trillionth of a nanosecond. Heavier particles last even less than that, as they require more energy to materialize. Now, what Tryon was suggesting was that our entire universe, with its vast amount of matter, was a huge quantum fluctuation, which somehow failed to disappear for more than 10 billion years. Everybody thought that was a very funny joke.

But Tryon was not joking. He was devastated by the reaction of his colleagues, to the extent that he forgot his idea and suppressed the memory of the whole incident. But the idea continued brewing at the back of his mind and resurfaced three years later. At that time, Tryon decided to publish it. His paper appeared in 1973 in the British science journal *Nature*, under the title "Is the Universe a Vacuum Fluctuation?"

Tryon's proposal relied upon a well-known mathematical fact—that the energy of a closed universe is always equal to zero. The energy of matter is positive, the gravitational energy is negative, and it turns out that in a closed universe the two contributions exactly cancel each other. Thus, if a closed universe were to arise as a quantum fluctuation, there would be no need to borrow energy from the vacuum and the lifetime of the fluctuation could be arbitrarily long.

The creation of a closed universe out of the vacuum is illustrated in Figure 17.3. A region of flat space begins to swell, taking the shape of a balloon. At the same time, a colossal number of particles are spontaneously created in that region. The balloon eventually pinches off, and—voilà—we have a closed universe, filled with matter, that is completely disconnected from the original space.[3] Tryon suggested that our universe could have originated in this way and emphasized that such a creation event would not re-

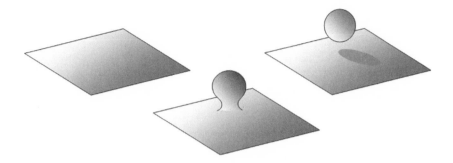

Figure 17.3. A closed universe pinches off a large region of space.

quire a cause. "In answer to the question of why it happened," he wrote, "I offer the modest proposal that our universe is simply one of those things which happen from time to time."[4]

The main problem with Tryon's idea is that it does not explain why the universe is so large. Closed baby universes are constantly pinched off any large region of space, but all this activity occurs at the Planck distance scale, as in the spacetime foam picture shown in Figure 12.1. Formation of a large closed universe is possible in principle, but the probability for this to happen is much smaller than that for a monkey to randomly type the full text of Shakespeare's *Hamlet*.

In his paper Tryon argued that even if most of the universes are tiny, observers can only evolve in a large universe and therefore we should not be surprised that we live in one. But this falls short of resolving the difficulty, because our universe is much larger than necessary for the evolution of life.

A more fundamental problem is that Tryon's scenario does not really explain the origin of the universe. A quantum fluctuation of the vacuum assumes that there was a vacuum of some pre-existing space. And we now know that "vacuum" is very different from "nothing." Vacuum, or empty space, has energy and tension, it can bend and warp, so it is unquestionably *something*.[5] As Alan Guth wrote, "In this context, a proposal that the universe was created from empty space is no more fundamental than a proposal that the universe was spawned by a piece of rubber. It might be true, but one would still want to ask where the piece of rubber came from."[6]

The picture of quantum tunneling from nothing has none of these problems. The universe is tiny right after tunneling, but it is filled with a false vacuum and immediately starts to inflate. In a fraction of a second, it blows up to a gigantic size.

Prior to the tunneling, no space or time exists, so the question of what happened *before* is meaningless. *Nothing*—a state with no matter, no space, and no time—appears to be the only satisfactory starting point for the creation.

•

A few years after publishing my tunneling-from-nothing paper, I realized that I had missed an important reference. Normally, one finds out about such things much sooner, by way of pesky e-mails from the omitted authors. But this author did not write to me, and for a good reason: he did his work more than 1,500 years ago. He was Saint Augustine, the bishop of Hippo, one of the major cities in northern Africa.

Augustine grappled with the question of what God was doing before the creation—a quest he eloquently described in his *Confessions*. "For if he was idle . . . and doing nothing, then why did he not continue in that state forever—doing nothing, as he has always done?" Augustine thought that in order to answer his question, he first had to figure out what time is: "What then is time? If no one asks me, I know what it is. If I wish to explain it to him who asks, I do not know." A lucid analysis led him to the realization that time could be defined only through motion and could not, therefore, exist before the universe. Augustine's final conclusion was that "[t]he world was made not in time, but simultaneously with time. There was no time before the world." And thus it is meaningless to ask what God was doing then. "If there was no time, there was no 'then.'"[7] This is very close to what I argued in my tunneling-from-nothing scenario.

I learned about Augustine's ideas accidentally, in a conversation with my Tufts colleague Kathryn McCarthy. I read the *Confessions* and quoted Saint Augustine in my next paper.[8]

MANY WORLDS

The universe emerging from quantum tunneling does not have to be perfectly spherical. It can have a variety of different shapes, and it can also be filled with different kinds of false vacuum. As usual in quantum theory, we cannot tell which of these possibilities has been realized, but can only calculate their probabilities. Could it be, then, that there is a multitude of other universes that started differently from our own?

This issue is closely related to the thorny question of how quantum probabilities are to be interpreted. As we discussed in Chapter 11, there are two major alternatives. According to the Copenhagen interpretation, quantum mechanics assigns probabilities to all possible outcomes of an experiment, but only one of these outcomes actually happens. The Everett interpretation, on the other hand, asserts that all possible outcomes are realized in disconnected, "parallel" universes.

If the Copenhagen interpretation is adopted, then the creation was a one-shot event, with a single universe popping out of nothing. This, however, leads to a problem. The most likely thing to pop out of nothing is a tiny Planck-sized universe, which would not tunnel, but would instantly recollapse and disappear. Tunneling to a larger size has a small probability and therefore requires a large number of trials. It appears to be consistent only with the Everett interpretation.

In the Everett picture, there is an ensemble of universes with all possible initial states. Most of them are Planck-sized "flicker" universes, which blink in and out of existence. But in addition, there are some universes that tunnel to a larger size and inflate. The crucial difference from the Copenhagen interpretation is that all these universes are not merely possible, but real.[9] Since observers cannot evolve in the "flickers," only large universes will be observed.

All of the universes in the ensemble are completely disconnected from one another. Each has its own space and its own time. Calculations show that the most probable—and thus the most numerous—of the tunneling universes are the ones nucleating with the smallest initial radius and the highest energy density of the false vacuum. Our best guess, then, is that our own universe also nucleated in this way.

In scalar field models of inflation, the highest vacuum energy density is reached at the top of the energy hill, and thus most of the universes will nucleate having the scalar field in that vicinity. This is the most favorable starting point for inflation. Remember, I promised to explain how the field got to the top of the hill. In the tunneling-from-nothing scenario, this is where it was when the universe came into being.

The nucleation of the universe is basically a quantum fluctuation, and its probability decreases rapidly with the volume it encompasses. Universes nucleating with a larger initial radius have smaller probability, and in the limit of infinite radius, the probability vanishes. An infinite, open universe has a strictly zero probability of nucleating, and thus all universes in the ensemble must be closed.

THE HAWKING FACTOR

In July 1983 several hundred physicists from all over the world gathered in the Italian city of Padova for the tenth International Conference on General Relativity and Gravitation. The conference site was the thirteenth-century Palazzo della Ragione, the old courthouse, located at the very heart of Padova. The ground floor of the Palazzo is taken by the famous food market, which spills outdoors into the adjacent piazza. The upper floor is occupied by a spacious hall, frescoed with signs of the zodiac around its perimeter. That's where the lectures were held. The highlight of the program was the talk by Stephen Hawking, entitled "The Quantum State of the Universe." The entrance to the lecture hall was through a long stairway, and it was a nontrivial task to carry Hawking in his wheelchair up the stairs. I was glad I arrived early, since by the time Hawking appeared on stage, the hall was completely packed.

In his talk Hawking unveiled a new vision for the quantum origin of the universe, based on the work he had done with James Hartle of the University of California at Santa Barbara.[10] Instead of focusing on the early moments of creation, he asked a more general question: How can we calculate the quantum probability for the universe to be in a certain state? The universe could follow a large number of possible histories before it got to that state, and the rules of quantum mechanics can be used to determine how much each particular history contributes to the probability.* The final result for the probability depends on what class of histories is included in the calculation. The proposal of Hartle and Hawking was to include only histories represented by spacetimes that have no boundaries in the past.

A space without boundaries is easy to understand: it simply means a closed universe. But Hartle and Hawking required that the spacetime should also have no boundary, or edge, in the past direction of time. It should be closed in all four dimensions, except for the boundary corresponding to the present moment (see Figure 17.4).

A boundary in space would mean that there is something beyond the universe, so that things can come in and go out through the boundary. A boundary in time would correspond to the beginning of the universe, where some initial conditions would have to be specified. The proposal of Hartle and Hawking asserts that the universe has no such boundary; it is

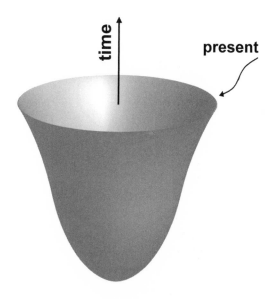

Figure 17.4. A two-dimensional spacetime without a past boundary.

*More precisely, the quantity called the *wave function* is found by adding up the contributions of different histories. The probability is given by the square of the wave function.

"completely self-contained and not affected by anything outside itself." That sounded like a very simple and attractive idea. The only problem was that spacetimes closed to the past, of the kind shown in Figure 17.4, do not exist. There should be three spacelike and one timelike direction at every spacetime point, but a closed spacetime necessarily has some pathological points with more than one timelike direction (see Figure 17.5).

To resolve this difficulty, Hartle and Hawking suggested that we switch from real time to Euclidean time. As we discussed earlier in this chapter, Euclidean time is no different from another spatial direction; so the spacetime simply becomes a four-dimensional space, and there is no problem making it closed. Thus, the proposal was that we calculate probabilities by adding up contributions from all Euclidean spacetimes without boundaries. Hawking emphasized that this was only a proposal. He had no proof that it was correct, and the only way to find out was to check whether or not it makes reasonable predictions.

The Hartle-Hawking proposal has a certain mathematical beauty about it, but I thought that after switching to Euclidean time it lost much of its in-

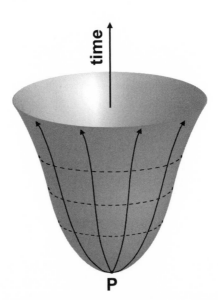

Figure 17.5. Same as in Figure 17.4, with timelike and spacelike directions indicated by solid and dashed lines, respectively. The point *P* is pathological, since all directions are timelike at that point.

tuitive appeal. Instead of summing over possible histories of the universe, it instructs us to sum over histories that are certainly impossible, because we do not live in Euclidean time. So, after the scaffolding of the original motivation is dropped, we are left with a rather formal prescription for calculating probabilities.[11]

In the conclusion of his talk, Hawking discussed the implications of the new proposal for the inflationary universe. He argued that the main contribution to the sum over histories is given by the Euclidean spacetime having the form of a hemisphere—the same as appeared in my tunneling calculation—and that the following evolution is represented by the inflationary expansion in ordinary time. (Switching back to ordinary time from the Euclidean formalism was a tricky procedure, which I will not try to describe here.) The result was the same spacetime history as in my Figure 17.3, but obtained from a very different starting point.

I expected that Hawking would mention my work on quantum tunneling from nothing and was disappointed when he didn't. But I was sure that with Hawking now in the field, the whole subject of quantum cosmology, and my work in particular, would receive a lot more attention than before.

MUCH ADO ABOUT NOTHING

An important difference between the "tunneling from nothing" and "no boundary" proposals is that they give very different, and in some sense opposite, predictions for the probabilities. The tunneling proposal favors nucleation with the highest vacuum energy and the smallest size of the universe. The no-boundary prescription, on the contrary, suggests that the most likely starting point is a universe of the smallest vacuum energy and largest possible size. The most probable thing to pop out of nothing is then an infinite, empty, flat space. I find this very hard to believe!

The conflict between the two approaches became apparent only after some initial confusion. The result in my 1982 paper was that larger universes had a *higher* probability of nucleating, so it looked as though the two proposals were in agreement. I kept returning to my calculation, because the result was so counter-intuitive. In 1984 I found an error, which reversed the probability trend. At the time, Hawking was visiting Harvard, and I rushed to talk to him and share my new insight. But Stephen was unconvinced and thought that I had gotten it right the first time around.*

Hawking is a legend among physicists and far beyond. I admire both his science and his spirit and treasure the opportunities to talk to him. Since it takes him so much effort to communicate, people are often reluctant to approach him. It took me a while to realize that Stephen actually enjoys conversation and does not even mind some joking around. We have very different views on eternal inflation and on quantum cosmology, but this only makes the discussion more interesting.

In 1988 I took the battle to Hawking's turf and gave a talk to his group at Cambridge University, highlighting the advantages of my approach. After the talk, Hawking rolled up to me in his wheelchair. I expected some critical remarks, but instead he invited me over for dinner. After a meal of duck with potatoes and a plum pie, cooked by Stephen's mother, we talked about the use of wormholes—shortcut tunnels through spacetime—for intergalactic travel. This is the physicist's notion of a light after-dinner conversation. As for the no-boundary proposal, Stephen did not change his mind.

The dispute between proponents of the two approaches is still going on. There was even an "official" debate at the Cosmo-98 conference in Monterey, California, with Hawking defending the no-boundary proposal and Andrei Linde and me arguing for the tunneling one.† It was not actually much of a debate. It takes a long time for Hawking to compose sentences with his speech synthesizer, so we did not progress much beyond the prepared statements.

We could resolve the dispute if we devised some observational test to distinguish between the two proposals. This, however, appears rather unlikely, and the reason is eternal inflation. Quantum cosmology makes predictions about the initial state of the universe, but in the course of eternal inflation any effect of the initial conditions is completely erased. Take, for example, the string theory landscape that we discussed before. We can start in one inflating vacuum or another, but inevitably bubbles of other vacua

---

*The error in my original paper was independently noticed and corrected by Andrei Linde, by Valery Rubakov, and by Yakov Zel'dovich and Alexei Starobinsky.

†The following day Hawking had another important engagement: he went to Hollywood to have his electronic voice recorded for a special episode of the animated television show *The Simpsons*.

Figure 17.6. Discussing quantum cosmology with Hawking. From left to right: the author, Bill Unruh of the University of British Columbia, and Stephen Hawking (drinking tea with the help of his nurse). (Courtesy of Anna Zytkow)

will be formed, and the entire landscape will be explored. The properties of the resulting multiverse will be independent of how inflation started.[12]

Thus, quantum cosmology is not about to become an observational science. The dispute between different approaches will probably be resolved by theoretical considerations, not by observational data. For example, the quantum state of the universe may be determined by some new, yet to be discovered, principle of string theory. It may, of course, differ from either of the present proposals. This issue is not likely to be settled any time soon.

18

The End of the World

Some say the world will end in fire,
Some say in ice.

—Robert Frost, "Fire and Ice"

M y account of the state of the universe would be incomplete without a description of how the world is going to end. The theory of inflation tells us that the universe as a whole will go on forever, but our local region—the observable universe—may well come to an end. This issue was at the center of cosmological research for a good part of the last century, and our picture of the end has changed several times in the process. I will now review the recent history of the subject and then give you the latest on cosmic eschatology.

GRIM OPTIONS

After Einstein denounced the cosmological constant early in the 1930s, the predictions of Friedmann's homogeneous and isotropic models were clear and simple: the universe will collapse to a big crunch if its density is greater than critical and will continue expanding forever otherwise. To determine the fate of the universe, all we had to do was to accurately measure the average density of matter and see whether it is greater than critical. If it is, then the expansion of the universe will gradually slow down and will be followed by contraction. Slow at first, the contraction will accelerate. Galaxies will get closer and closer, until they merge into a huge conglomerate of stars. The sky will get brighter, not because of the stars—they will most probably all be dead by then—but because of the increased intensity of the cosmic background radiation. The radiation will heat up the remnants of stars and planets to very uncomfortable temperatures, and any living creatures who managed to survive until then will end their days like lobsters in boiling water.

The stars will eventually be disrupted in collisions with one another or get vaporized by the intense heat of the radiation. The resulting hot fireball will be similar to the one in the early universe, except now it will be contracting rather than expanding. Another difference from the big bang is that the contracting fireball is rather inhomogeneous. Denser regions collapse first to form black holes, which then merge into larger black holes, until they all merge together at the big crunch.

In the opposite case of less-than-critical density, the gravitational pull of matter is too weak to turn the expansion around. The universe will then expand forever. In less than a trillion years all stars will exhaust their nuclear fuel. Galaxies will turn into swarms of cold stellar remnants—white dwarfs, neutron stars, and black holes. The universe will be completely dark, with ghostly galaxies flying apart into the expanding void.

This state of affairs endures for at least 10^{31} years, but eventually nucleons that make up the stellar remnants decay, turning into lighter particles—positrons, electrons, and neutrinos. Electrons and positrons annihilate into photons, and the dead stars begin slowly to dissolve. Even black holes do not last forever. Hawking's famous insight, that a black hole leaks out quanta of radiation, implies that it gradually loses all its mass, or, as physicists say, "evaporates." One way or another, in less than a google years all familiar structures in the universe will be gone. Stars, galaxies, and clusters will disappear without a trace, leaving behind an ever-thinning mix of neutrinos and radiation.[1]

The fate of the universe is encoded in the parameter called Omega, defined as the average density of the universe divided by the critical density. If Omega is greater than 1, the universe will end in fire and a big crunch; if it is less than 1, we can look forward to freezing and slow disintegration. In the borderline case of Omega equal to 1, the expansion gets increasingly slow, but never stops completely. The universe narrowly escapes the big crunch, only to become a frozen graveyard.

For more than half a century astronomers worked very hard trying to measure the value of Omega. However, nature was not in the mood to disclose her long-term plans. Omega was tantalizingly close to 1, but the accuracy of measurements was not sufficient to tell whether it was above or below.

INFLATIONARY TWIST

Our view of the end was transformed in the 1980s, when the idea of inflation appeared on the scene. Prior to that, a big crunch and an unlimited expansion seemed equally probable, but now the theory of inflation made a very definite prediction.

During inflation the density of the universe is driven extremely close to the critical density. Depending on quantum fluctuations of the scalar field, some regions have density above and other regions below critical, but on average the density is almost exactly critical. Thus, anyone who had nightmares that the universe might collapse to a big crunch in a few trillion years could now relax. The end will be slow and unexciting, as the cold remnant of our Sun hangs around for eons, waiting for all of its nucleons to decay.

A characteristic feature of the critical-density universe is that the structure formation process is stretched over an enormous length of time, with larger structures taking a longer time to assemble. Galaxies form first, they later clump into clusters, and still later the clusters clump into superclusters. If the average density in our observable region is above critical, then in about a hundred trillion years this entire region will turn into a huge super-duper cluster. By that time all stars will already be dead and all observers will probably be extinct, but structure formation will continue, extending to larger and larger scales. It will only stop when the cosmic structures disintegrate as a result of nucleon decay and black hole evaporation.

Another twist introduced by inflation is that the entire universe will never come to an end. Inflation is eternal. Countless regions similar to ours will be formed in other parts of the inflating spacetime, and their inhabitants will struggle to understand how it all began and how it will end.

GALACTIC SOLITUDE

Friedmann's relation between the density of the universe and its ultimate fate applies only if the vacuum energy density (the cosmological constant) is equal to zero. This was the standard assumption before 1998, but when evidence to the contrary was discovered, all earlier forecasts for the future of the universe had to be revised. The main prediction that the world will end (locally) in ice rather than in fire did not change, but some details had to be modified.

As we discussed earlier, the expansion of the universe begins to accelerate once the density of matter drops below that of the vacuum. Any gravitational clumping stops at that time. Clusters of galaxies that are already bound together by gravity survive, but looser groups are dispersed by the repulsive gravity of the vacuum.

Our Milky Way is bound to the *Local Group*, which includes the giant spiral of Andromeda and some twenty dwarf galaxies. Andromeda is on a collision course with the Milky Way; they are going to merge in about 100 billion years. Galaxies beyond the Local Group will all speed away, moving faster and faster. One by one, they will cross our horizon and disappear from view. This process will be complete a few hundred billion years from now. In that remote epoch, astronomy will become a very boring subject. Apart from the giant galaxy resulting from our union with Andromeda and its dwarf satellites, the sky will be completely empty.[2] We should enjoy the show while it lasts!

THE FINAL VERDICT

Our forecast for the universe would now be complete if the cosmological constant were truly a constant. But as we know, there are good reasons to believe that the vacuum energy density varies in a very wide range, taking different values in different parts of the universe. In some regions it is large and positive, in other regions large and negative, but only in rare regions where it is close to zero will there be some creatures who care about what it is.

It follows that the value we observe here is not the lowest possible energy density; inevitably, it will get smaller in the future. Consider, for example, Linde's model, where the vacuum energy is due to a scalar field with a very gently sloping energy landscape (Figure 13.1). The slope is so small that the field changed very little in the 14 billion years since the big bang. But eventually it will start rolling downhill, and cosmic acceleration will begin slowing down. At some point the field will get below the zero level, to the negative values of energy density. A negative-energy vacuum is gravitationally attractive, so before long cosmic expansion will stop and contraction will begin.

An alternative scenario, suggested by the string theory landscape picture, is that classically our vacuum is stable and has a constant energy density, but quantum-mechanically it can decay through bubble nucleation. Bubbles of negative-energy vacuum will occasionally pop out and expand at a speed quickly approaching the speed of light. A bubble wall may be charging toward us at this very moment. We are not going to see it coming: the light will not get much ahead of the wall, because it is moving so fast. Once the wall hits, our world will be completely annihilated. The particles that make up stars, planets, and our bodies will not even exist in the new vacuum. All the familiar objects will be instantly destroyed and turned into clumps of some alien forms of matter.

One way or the other, the vacuum energy will eventually turn negative in our local region. The region will then start contracting and will collapse to a big crunch.[3] Exactly when this is going to happen is hard to predict. The rate of bubble nucleation can be extremely low, and it may take googles of years for our neighborhood to be hit by a bubble wall. In scalar field models, the time of apocalypse depends on the slope of the energy hill and may come as soon as in 20 billion years.

· 19 ·

Fire in the Equations

What is it that breathes fire into the equations and makes a
universe for them to describe?   —STEPHEN HAWKING

ALFONSO'S ADVICE

Alfonso the Wise, the thirteenth-century king of Castile, had a great respect for astronomy. And for a very practical reason: the knowledge of precise locations of planets on the sky was vital for casting accurate horoscopes. To improve their accuracy, Alfonso commissioned new astronomical tables based on Ptolemy's model of the universe—then the latest word in cosmology. But when the intricacies of Ptolemy's system were explained to him, Alfonso was rather skeptical: "If the Lord Almighty had consulted me before embarking upon creation, I should have recommended something simpler."[1]

King Alfonso could have made a similar remark about the worldview that I have described in this book. It asserts the existence of an infinite ensemble of universes, each containing a tapestry of regions with different particle physics. Regions where intelligent creatures can exist are rare and are separated by enormous distances. Rarer still are regions that are completely identical to one another, and yet there is an infinity of such regions. How wasteful of space, matter, and universes!

However, the number of universes is not something to be unduly concerned about. The new worldview saves a more precious commodity: it greatly reduces the number of arbitrary assumptions we have to make about the universe. The best theory is the one that explains the world with the fewest and simplest assumptions.

Earlier cosmological models suggested a Creator meticulously designing and fine-tuning the universe. Every detail of particle physics, each constant of nature, and all the primordial ripples had to be set just right. One can imagine the volumes and volumes of specifications the Creator handed down to his assistants to complete the job! The new worldview evokes a different image of the Creator. After some thought, he comes up with a set of equations for the fundamental theory of nature. This initiates the process of runaway creation. No further instructions are needed: the theory describes quantum nucleation of universes from nothing, the process of eternal inflation, and the creation of regions with every possible type of particle physics, ad infinitum. Any specific member of this ensemble of universes is incredibly complicated and would take an enormous amount of information to describe. But the entire ensemble can be codified in a relatively simple set of equations.[2]

GOD AS A MATHEMATICIAN

How do we know which portrait of the Creator is closer to the truth? Did he strive to optimize the use of "resources" like space and matter, or was he more concerned about having a concise mathematical description of nature? Unfortunately, he does not give interviews, but his product—the universe—leaves little doubt about what kind of Creator he is.

A casual look at the universe shows that space and matter are wasted in it with great abandon. Countless galaxies are scattered over immense stretches of nearly empty space. The galaxies fall into a few different classes, like spiral and elliptical, dwarf and giant. But apart from that, they are very similar to one another. The Creator makes it very clear that he is not embarrassed to repeat himself endlessly.

A more detailed examination reveals that the Creator is obsessed with mathematics. Pythagoras, in the sixth century B.C., was probably the first to suggest that mathematical relations were at the heart of all physical phenomena. His insight was confirmed by centuries of scientific research, and we now take it for granted that nature follows precise mathematical laws. But if you stop to think about it, this fact is highly peculiar.

Mathematics appears to be a product of pure thought, with a very loose relation to experience. But then how come it is so ideally suited to describing the physical universe? This is what the physicist Eugene Wigner called "the unreasonable effectiveness of mathematics in natural sciences." Consider the *ellipse* as a simple example. It was known to the ancient Greeks as the curve you get by cutting a cone with a plane at an angle. Archimedes and other Greek mathematicians studied the properties of the ellipse out of sheer interest in geometry. Then, almost two thousand years later, Johannes Kepler discovered that planets in their motion around the Sun describe ellipses with a remarkable accuracy. But what does the motion of Venus and Mars have to do with sections of a cone?

Closer to home, in the 1960s my mathematician friend Victor Kac investigated a class of intricate mathematical structures now known as Kac-Moody algebras. His only motivation was his nose, which told him that the structures smelled interesting and could yield some beautiful mathematics. No one could have predicted that in a couple of decades these algebras would play a major role in string theory.

These examples are not exceptions. More often than not, physicists discover that the mathematics they need to describe a new class of phenomena has already been studied by mathematicians, for reasons that have nothing to do with the phenomena in question. It appears that the Creator shares the mathematicians' sense of beauty. Many physicists rely on his idiosyncrasy and use mathematical beauty as a guide in their search for new theories. According to Paul Dirac, one of the pioneers of quantum mechanics, "It is more important to have beauty in one's equations than to have them fit experiment . . . because the discrepancy may be due to minor features . . . that will get cleared up with further development of the theory."[3]

Mathematical beauty is no easier to define than beauty in art.[4] An example of what mathematicians find beautiful is what is known as *Euler's formula*, $e^{i\pi} + 1 = 0$. One criterion for beauty is simplicity, but simplicity alone does not do it. The relation $1 + 1 = 2$ is simple, but not particularly beautiful because it is trivial. In contrast, Euler's formula shows a rather surprising connection between three seemingly unrelated numbers: the number e, which is related to "natural" logarithms; the "imaginary" number i—the square root of -1; and the number π—the ratio of the circumference of a circle to its diameter. We can call this property "depth." Beautiful mathematics combines simplicity with depth.[5]

If indeed the Creator has a mathematician's mind, then the equations of the fundamental theory of nature should be wonderfully simple and unbelievably deep. Some people think that this final theory is the theory of strings, which we are now in the process of discovering. This theory is definitely very deep. It does not look simple, but simplicity may emerge when the theory is better understood.

MATHEMATICAL DEMOCRACY

If we ever discover the final theory of nature, the question will still remain: Why this theory? Mathematical beauty may be useful as a guide, but it is hard to imagine that it will suffice to select a unique theory out of the infinite number of possibilities. As the physicist Max Tegmark put it, "Why should one mathematical structure, and only one, out of all the countless mathematical structures, be endowed with physical existence?"[6] Tegmark, now at Massachusetts Institute of Technology, suggested a possible way out of this impasse.

His proposal is as simple as it is radical: he argues that there should be a universe corresponding to each and every mathematical structure.[7] There is, for example, a Newtonian universe governed by the laws of Euclidean geometry, classical mechanics, and Newton's theory of gravitation. There are also universes where space has an infinite number of dimensions, and others having two dimensions of time. Even harder to imagine is a universe governed by the algebra of quaternions,* which has neither space nor time.

Tegmark asserts that all these universes exist "out there." We are not aware of them, just as we are not aware of other universes nucleating out of nothing. The mathematical structures in some of the universes are intricate enough to allow the emergence of "self-aware substructures" like you and

---

*The quaternion is a generalization of the more familiar complex number. It has one real and three imaginary parts.

me. Such universes are rare, but of course they are the only ones that can be observed.

We have no evidence to support this dramatic extension of reality. The only reason for elevating universes with other mathematical structures to the realm of existence is to avoid explaining why they do not exist. This may be enough to convince some philosophers, but physicists need something more substantial. In the spirit of the principle of mediocrity, one could try to show that the fundamental theory of our universe is in some sense typical of all the theories rich enough to harbor observers. This would lend support to Tegmark's extended multiverse.

If successful, this line of reasoning would drive the Creator entirely out of the picture. Inflation relieved him of the job of setting up the initial conditions of the big bang, quantum cosmology unburdened him of the task of creating space and time and starting up inflation, and now he is being evicted from his last refuge—the choice of the fundamental theory of nature.

Tegmark's proposal, however, faces a formidable problem. The number of mathematical structures increases with increasing complexity, suggesting that "typical" structures should be horrendously large and cumbersome. This seems to be in conflict with the simplicity and beauty of the theories describing our world.[8] It thus appears that the Creator's job security is in no immediate danger.

MANY WORLDS IN ONE

Philosophers and theologians have been arguing for centuries, trying to decide whether the universe is finite or infinite, stationary or evolving, eternal or transient. You might have thought that all possible answers to these questions have already been anticipated. However, the worldview emerging from recent developments in cosmology is not what anyone had expected. Instead of choosing between conflicting options, it suggests that each of them has some element of truth.

At the heart of the new worldview is the picture of an eternally inflating universe. It consists of isolated "island universes," where inflation has ended, immersed in the inflating sea of false vacuum. The boundaries of these postinflationary islands are rapidly expanding, but the gaps that separate them are widening even faster. Thus there is always room for more island universes to form, and their number increases without bound.

Viewed from the inside, each island is a self-contained infinite universe. We live in one of these island universes, and our observable region is one of the infinite number of O-regions that it contains. It is conceivable that billions of years from now our descendants will travel to other O-regions, but a voyage to another island universe is impossible, even in principle. No matter how long we travel and how fast, we are forever confined to our own island universe.

The entire eternally inflating spacetime originated as a minuscule closed universe. It tunneled, quantum-mechanically, out of nothing and immediately plunged into the never-ending fury of inflation. Thus the universe is eternal, but it did have a beginning.

Inflation rapidly blows the universe up to an enormous size, but from a global viewpoint it always remains closed and finite. And yet, due to the peculiar structure of inflationary spacetime, it contains an unlimited number of infinite island universes.

Constants of nature that shape the character of our world take different values in other island universes. Most of these universes are drastically different from ours, and only a tiny fraction of them are hospitable to life.[9] Observers in each habitable island will find that their universe evolves from a big bang to a big crunch. However, in the global view, all types of islands at all stages of their evolution are present simultaneously. This situation is analogous to that of the human population of the Earth. Each person starts as a baby and grows older with time, but the entire population includes people of all ages at any given moment. Although the total volume of the universe grows with time, the fraction of space occupied by each type of island does not change. In this sense, the eternally inflating universe is stationary.

A striking feature of the new worldview is the existence of multiple "other worlds" beyond our observable region. Some of them are rather uncontroversial. Very few people, for example, would question the reality of other O-regions, even though they cannot be observed. We do have some circumstantial evidence for multiple island universes with diverse properties. As for the other, disconnected spacetimes that nucleated out of nothing, we have no idea how to test their existence observationally.

The picture of quantum tunneling from nothing raises another intriguing question. The tunneling process is governed by the same fundamental laws that describe the subsequent evolution of the universe. It follows that the laws should be "there" even prior to the universe itself. Does this mean that the laws are not mere descriptions of reality and can have an independent existence of their own? In the absence of space, time, and matter, what tablets could they be written upon? The laws are expressed in the form of mathematical equations. If the medium of mathematics is the mind, does this mean that mind should predate the universe?

This takes us far into the unknown, all the way to the abyss of great mystery. It is hard to imagine how we can ever get past this point. But as before, that may just reflect the limits of our imagination.

Epilogue

To: Galactic Council
From: WSX-23EDJ

Greetings! As required by the Protocol, I have completed my inspection of the planet Earth, located in sector S-16 in the peripheral zone of the Galaxy. The human race populating this planet has made good progress in the 1000 Earth years since the last inspection. I have upgraded their status from "budding" to "technologically challenged."

You will be amused to know that humans believe they are close to discovering the final theory of the universe. I envy their youthful enthusiasm . . . On certain issues they have come close to the right answers—surprisingly, I would say, for a primitive civilization such as this. In other matters, though, they are pretty far behind. They have not even figured out the right questions.

Overall, this race is still rather immature. I recommend against inclusion in the Galactic Union at this time. Further details will be forthcoming in my regular report.

Yours respectfully,
WSX-23EDJ

Notes

1. WHAT BANGED, HOW IT BANGED, AND WHAT CAUSED IT TO BANG

1. A. H. Guth, *The Inflationary Universe* (Addison-Wesley, Reading [Mass.], 1997, p. 2).

2. THE RISE AND FALL OF REPULSIVE GRAVITY

1. Einstein to Ehrenfest, January 16, 1916 (as quoted in A. Pais, *Subtle is the Lord* (Oxford University Press, Oxford, 1982).
2. Einstein to Sommerfeld, February 8, 1916, ibid.
3. It was later realized that Einstein's static cosmological model is not acceptable even on purely theoretical grounds, because the balance between attractive and repulsive gravity in this model is unstable. If for some reason the size of the universe is slightly increased, the matter density will go down (since the distances between galaxies will grow), while the vacuum energy density will remain the same, being fixed by the cosmological constant. Hence, the repulsive gravity of the vacuum will now be stronger than the attractive gravity of matter and will cause the universe to expand. This will lead to a further increase of volume and to even greater imbalance between the attractive and repulsive forces. The universe will thus enter a regime of runaway expansion. Similarly, if the size of Einstein's static universe is slightly decreased, the attractive gravity of matter will win over the repulsion of the vacuum and the universe will collapse to a point. Small fluctuations in the size of the universe are inevitable according to the quantum theory, and thus Einstein's universe cannot remain in balance for an infinite time.

3. CREATION AND ITS DISCONTENTS

1. As quoted in E. A. Tropp, V. Y. Frenkel, and A. D. Chernin, *Aleksandr Aleksandrovich Fridman* (Nauka, Moscow, 1988, p. 133).

2. Friedmann did not consider the case of a spatially flat universe. It was studied by Einstein and Willem de Sitter in 1932.

3. A notable exception was Einstein's reaction to Friedmann's work. Initially, Einstein thought that Friedmann had made a mistake and wrote a brief note to the journal pointing to what he thought was an error. However, in less than a year he had to withdraw his criticism after a conversation with Friedmann's friend, Yuri Krutkov. Krutkov reported home that he had won a debate with Einstein and that "Petrograd's honor is saved!" But Einstein, though he agreed with Friedmann's mathematics, still believed that the universe was static and that Friedmann's work was therefore of purely formal interest. In his second note to the journal, he wrote that he was "convinced that Mr. Friedmann's results are both correct and clarifying." He added in the original draft that the results could hardly be of any physical significance, but then crossed this phrase out, perhaps realizing that it was based more on his philosophical prejudice than on any known fact. Quotes are from Helge Kragh, *Cosmology and Controversy*, (Princeton University Press, Princeton [N.J.], 1996).

4. The source of stellar energy was not known in Helmholtz's time, but now we know that stars are burning nuclear fuel by turning hydrogen into helium and then into heavier nuclei. This is an irreversible process accompanied by an increase of entropy, and eventually stars run out of nuclear fuel. Some stars turn off their nuclear engines without much fanfare and then gradually cool down, while others explode, throwing their constituent gas into the interstellar space and leaving behind a compact remnant (a neutron star or a black hole). The expelled gas can be reprocessed to form new stars, but sooner or later the gas supply will be exhausted, as more and more of it ends up in cold stellar remnants. In a trillion years from now, galaxies will probably be noticeably dimmer than they are today. The process of gradual dimming of lights may be rather protracted, but one thing is clear: the universe as we know it could not have existed forever.

5. Boltzmann's fluctuation idea is probably the first example of what will later be known as *anthropic arguments* (see Chapter 13).

6. The first persuasive evidence for the galactic evolution was presented in the 1950s by the Cambridge astronomer Martin Ryle. He found that powerful radio emission from galaxies was much more common a few billion years ago than it is now.

7. Arthur Conan Doyle, *The Sign of Four*.

4. THE MODERN STORY OF GENESIS

1. Quoted from R. H. Stuewer, in *The Kaleidoscope of Science*, ed. by E. Ullmann-Margalit (Reidel, Dordrecht [Netherlands], 1986, p. 147).

2. The description of Gamow's life in this section is based mostly on his unfinished autobiography, *My World Line* (Viking Press, New York, 1970).

3. Atoms are made of small, positively charged nuclei and negatively charged electrons "orbiting" around them. (I put "orbiting" in quotation marks, because quantum uncertainties are important in the atom, so instead of picturing electrons as moving in

an orderly way along their orbits, like planets around the Sun, it is more accurate to picture them as being "smeared" around the orbits.) The nuclei consist of two types of subatomic particles: protons, which have a positive electric charge, and neutrons, which are electrically neutral. The chemical properties of an atom are determined solely by the number of electrons (which is equal to the number of protons, so that the atom is electrically neutral).

4. The origin of this imbalance between matter and antimatter is one of the active areas of research in modern cosmology. For a discussion, see A. H. Guth, *The Inflationary Universe* (Addison-Wesley, Reading [Mass.], 1997).

5. A more detailed discussion of the hot fireball and element formation can be found in Steven Weinberg's classic bestseller *The First Three Minutes* (Bantam, New York, 1977).

6. M. J. Rees, *Before the Beginning* (Addison-Wesley, Reading [Mass.], 1997, p. 17).

7. S. Weinberg, op. cit., p. 123.

5. THE INFLATIONARY UNIVERSE

1. The twists and turns of Alan Guth's path to the discovery of inflation are described in his excellent book *The Inflationary Universe: The Quest for a New Theory of Cosmic Origins* (Addison-Wesley, Reading [Mass.], 1997).

2. It is conceivable that our vacuum is not, in fact, the lowest-energy one. String theory, which is now the prime candidate for the fundamental theory of nature, suggests the existence of negative-energy vacua. If they do exist, then our vacuum may eventually decay, with catastrophic consequences for all the material objects it contains. We shall discuss string theory in Chapter 15 and the possibility of our vacuum decay in Chapter 18. Until then we shall assume that we live in the true vacuum.

3. This conclusion is easy to understand from simple energy considerations. The force on a physical object always acts in the direction of reducing its energy (more precisely, its *potential* energy, that is, the part of energy not related to motion). For example, the force of gravity pulls objects down, so that their energy is decreased. (The gravitational energy grows with elevation above the ground.) For a false vacuum, the energy is proportional to the volume it occupies and can be reduced only by reducing the volume. Hence, there should be a force causing the vacuum to shrink. This is the force of tension.

6. TOO GOOD TO BE WRONG

1. A. H. Guth, "The inflationary universe: A possible solution to the horizon and flatness problems," *Physical Review*, vol. D23, p. 347 (1981).

2. The Starobinsky model is based on a modified form of Einstein's gravitational equations. The modification becomes important only when the curvature of spacetime gets very high. The magnitude of the curvature plays the role of a scalar field in this theory.

3. True to the Russian style, Mukhanov and Chibisov wrote their paper "for Landau,"

stating their result and providing little detail of how it was derived. Some of the Nuffield participants argue that an important step may be missing in this derivation and that Mukhanov and Chibisov may not, therefore, deserve full credit for the result. I will not attempt to settle this issue here.

8. RUNAWAY INFLATION

1. A. Vilenkin, "The birth of inflationary universes," *Physical Review*, vol. D27, p. 2848 (1983). This paper is about quantum cosmology; eternal inflation is discussed in sections IV and V.

2. An exponentially expanding region would quickly cover the whole computer screen, forcing us to stop the simulation. We dealt with this problem by using an expanding distance scale, which grew at the same rate as the inflating regions. Measured by this expanding ruler, the size of the inflating false-vacuum volume does not change in time, so it occupies a fixed area on the screen. In the economic inflation analogy that we used in Chapter 5, this method of measurement corresponds to expressing prices in "original dollars," so that the effect of inflation is factored out.

3. M. Aryal and A. Vilenkin, "The fractal dimension of the inflationary universe," *Physics Letters*, vol. B199, p. 351 (1987).

4. A. D. Linde, "Eternally existing self-reproducing chaotic inflationary universe," *Physics Letters*, vol. B175, p. 395 (1986). The term "eternal inflation" was introduced by Linde in this paper.

9. THE SKY HAS SPOKEN

1. The accelerated expansion of the universe was discovered by the High-Redshift Supernova Search Team, led by the Harvard astronomer Robert Kirshner and by Brian Schmidt of Siding Springs Observatory in Australia, and by the Supernova Cosmology Project team, led by Saul Perlmutter. For a witty firsthand account of this discovery, see Robert Kirshner's book *The Extravagant Universe: Exploding Stars, Dark Energy, and the Accelerating Cosmos* (Princeton University Press, Princeton, [N.J.], 2004).

2. Inflation can be reconciled with density smaller than critical at the expense of making the theory more complicated and less attractive. To this end, the scalar field energy landscape needs to be specially designed. It needs to have a barrier, as in Guth's original model (Figure 6.2). But instead of dropping steeply toward the minimum, the barrier must be followed by a very gentle slope. The resulting model combines the features of Guth's old inflation scenario with the improved scenario by Linde and others. The field tunnels through the barrier via bubble nucleation and completes its journey to the minimum by slowly rolling downhill within individual bubbles. In his analysis of vacuum bubbles, Sidney Coleman showed that from within they look like open Friedmann universes with density smaller than critical. By carefully adjusting the height and the slope of the hill, one can arrange for the density to be close, but not too

close, to the critical density. Physicists find such fine-tuning very distasteful, so the hope is that it will not be needed.

If, on the other hand, observations point to a density greater than critical, by more than one part in 100,000, the implication would be that the universe is a relatively small three-dimensional sphere, not much larger than the present horizon. This would pose a severe problem for inflation.

3. The origin of gravitational waves is similar to that of the density perturbations (see Chapter 6). They are produced as quantum fluctuations during inflation, with amplitudes independent of their length scale. The prediction of gravitational waves follows from the work that Alexei Starobinsky did in 1980, before Guth proposed the idea of inflation.

4. Clover will start operating in 2008. It will be able to detect gravitational waves from inflation only if the false vacuum had a grand-unification energy scale. For a lower-energy vacuum, a more sensitive instrument will be needed.

10. INFINITE ISLANDS

1. A. D. Linde, "Life after inflation," *Physics Letters*, vol. B211, p. 29 (1988).
2. In flat spacetime, the square of the interval between two events is defined as (time separation)^2 − (space separation)^2. Except for the minus sign, this quantity is very similar to the length squared in the Pythagorean theorem. To calculate the interval, time and space separations have to be expressed in compatible units. For example, if time is measured in years, then length should be measured in light-years. The interval is timelike if its square is positive, and is spacelike if it is negative. For the class reunion and superball events discussed in the text, the time separation is 3 years, the space separation is 4 light-years, so the interval squared is $3^2 - 4^2 = -7$. Hence, the interval is spacelike.

11. THE KING LIVES!

1. J. Garriga and A. Vilenkin, "Many worlds in one," *Physical Review*, vol. D64, p. 043511 (2001).
2. A. D. Sakharov, *Alarm and Hope* (Knopf, New York, 1978).
3. G.F.R. Ellis and G. B. Brundrit, "Life in the infinite universe," *Quarterly Journal of the Royal Astronomical Society*, vol. 20, p. 37 (1979).
4. For a thought-provoking discussion of the many-world interpretation, see the book by David Deutsch, *The Fabric of Reality* (Penguin, New York, 1997).
5. As quoted in G. Edelman, *Bright Air, Brilliant Fire: On the Matter of the Mind* (Penguin, New York, 1992, p. 216).
6. This formulation is David Mermin's; see *Physics Today*, April 1989, p. 9.
7. From President Clinton's testimony to the grand jury on August 17, 1998.
8. An example of energy landscape designed to avoid eternal inflation is shown in the figure on p. 214 (compare with Figure 6.4).

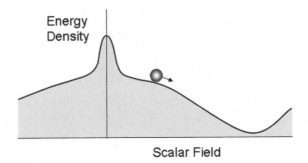

The flat hilltop responsible for eternal inflation is removed and replaced with a steep spike. At the same time, the flattened slope of the hill needs to be preserved, since otherwise we would have no inflation at all. Such landscapes are not likely to arise from particle physics. Inflation is eternal in practically all models suggested so far.

9. Some ethical implications of the new worldview are discussed in the paper I wrote with the philosopher Joshua Knobe and my Tufts colleague Ken Olum, "Philosophical implications of inflationary cosmology," to appear in the March 2006 issue of *The British Journal for the Philosophy of Science*.

12. THE COSMOLOGICAL CONSTANT PROBLEM

1. The first convincing measurement of electromagnetic vacuum fluctuations was performed only in the late 1990s, using the idea proposed decades earlier by the Dutch physicist Hendrik Casimir. Two metal plates are placed in a vacuum parallel to one another. Electromagnetic oscillations are suppressed in metal, and this has the effect of reducing vacuum fluctuations in the space between the plates. The pressure exerted by the fluctuating fields on the outer surfaces of the plates is therefore greater than the pressure acting from the inside, so there is a net force pushing the plates together. This force is very small and rapidly drops as the distance between the plates is increased. The measurement was performed for plates separated by about 1 micron (one-millionth of a meter).

2. This is exactly what happens in particle theories that have a special kind of symmetry, called *supersymmetry*. Bosons and fermions in such theories come in pairs, so that each Bose particle has a fermionic "partner" and vice versa. Partner particles in each pair have the same mass, and the vacuum energies of fermions and bosons exactly cancel one another. Hence, the total energy density of the vacuum is zero.

   This would be a neat solution to the cosmological constant problem, but the trouble is that our world is definitely not supersymmetric. Otherwise, we would see the partners of electrons, quarks, and photons copiously produced in particle accelerators. But none of these partner particles has ever been observed. Moreover, even in a supersymmetric world, the cancellation of the cosmological constant works only in the absence of gravity. The vacuum energy gets large and negative when gravity is taken into account.

13. ANTHROPIC FEUDS

1. C. J. Hogan, "Quarks, electrons and atoms in closely related universes," in *Universe or Multiverse*, ed. by B. J. Carr (Cambridge University Press, Cambridge, 2006).

2. Numerous examples of the apparent fine-tuning of the constants of nature are discussed in the article by Bernard J. Carr and Martin J. Rees in *Nature*, vol. 278, p. 605 (1979), and in the books *The Accidental Universe* (Cambridge University Press, Cambridge, 1982) by Paul C. W. Davies; *The Anthropic Cosmological Principle* (Oxford University Press, Oxford, 1986) by John D. Barrow and Frank J. Tipler; and *Universes* (Routledge, London, 1989) by John Leslie. For a lucid popular account, see Martin Rees's books *Before the Beginning: Our Universe and Others* (Addison-Wesley, Reading, 1997) and *Just Six Numbers* (Basic Books, New York, 2001).

3. B. Carter, "Large number coincidences and the anthropic principle in cosmology," in *Confrontation of Cosmological Theories with Observational Data*, ed. by M. S. Longair (Reidel, Boston, 1974, p. 132).

4. Stars less massive than the Sun have longer lifetimes. However, they tend to be unstable and are subject to flare-ups that can extinguish planetary life. We assume that planets orbiting such stars are disqualified as potential homes for observers.

5. Dicke presented this argument in 1961, in response to the intriguing hypothesis advanced by Paul Dirac, the famous British physicist. Dirac was struck by the weakness of gravity, which is 10^{40} times weaker than the electromagnetic force. He also noticed that the visible universe is 10^{40} times larger than the proton. Dirac thought this could not be a pure coincidence and suggested that the two numbers should somehow be connected. But the size of the visible universe grows with time, so its ratio to the size of the proton will be greater at later epochs. This led Dirac to conclude that the other number, expressing the weakness of gravity, should also grow: gravity must be getting progressively weaker.

Now, Dicke's argument gave a completely different perspective on the large-number coincidence. We are observing the universe not at some arbitrary epoch, but at the time when its age is comparable to the lifetime of a star. Dicke showed that at that particular time Dirac's large numbers are indeed close to one another. (This is not an accident: the visible universe is large because the stellar lifetimes are long, and long stellar lifetimes are in turn related to the weakness of gravity, thus establishing the connection between the two large numbers.) Thus the coincidence is automatically satisfied at the epoch when observers can exist, and there is no need to postulate any weakening of gravity. Precise astronomical measurements later showed that the strength of gravity remains constant with a very high accuracy. If there is any change, it must be smaller than 1 part in 10^{11} per year—much less than required by the Dirac hypothesis.

6. N. Bostrom, *Anthropic Bias* (Routledge, New York, 2002).

7. As quoted in A. L. Macay, A *Dictionary of Scientific Quotations* (Institute of Physics Publishing, Bristol [U.K.], 1991, p. 244).

8. David Gross, as quoted in "Zillions of universes? Or did ours get lucky?" by Dennis Overbye in *The New York Times*, October 28, 2003.

9. Paul Steinhardt, as quoted in "Out in the cold" by Marcus Chown in *New Scientist*, June 10, 2000.

14. MEDIOCRITY RAISED TO A PRINCIPLE

1. The doomsday argument is a fascinating and controversial subject. For a thought-provoking discussion, see *The End of the World* by John Leslie (Routledge, London, 1996) and *Time Travel in Einstein's Universe* by Richard Gott (Houghton Mifflin Company, Boston, 2001).

2. In an infinite universe, the volume factor can be defined as the fraction of volume occupied by regions of a given type. This definition, however, can lead to ambiguities. To illustrate the nature of the problem, consider the question, What fraction of all integers are odd? Even and odd integers alternate in the sequence 1,2,3,4,5, . . . , so you might think that the answer is obviously "half." The integers, however, can be ordered in a different way. For example, we could write 1,2,4,3,6,8, . . . This sequence still includes all integers, but now each odd integer is followed by two even ones; it appears that only a third of all integers are odd. The same sort of ambiguity arises in calculations of the volume factor in models of eternal inflation. Some interesting ideas have been proposed on how to deal with this difficulty, but at present the problem is still unresolved.

3. This is a bit of an oversimplification. Galaxies come in different sizes, from dwarfs to giants, with very different numbers of stars and, therefore, of observers. However, the vast majority of stars are in giant galaxies like ours. So the problem can be fixed by simply counting only giant galaxies and disregarding the rest.

   A more serious problem is that the density of matter and other characteristics of galaxies may change because of variation of life-neutral constants. For example, if the density perturbation parameter Q gets larger, galaxies form earlier and have a higher density of matter. As a result, close encounters between stars, which can disrupt planetary orbits and extinguish life, become more common. (This point was made by Max Tegmark and Martin Rees in their paper published in *Astrophysical Journal* in 1998.) Even if the encounter is not close enough to affect the planets, it may disturb the swarm of comets in the outer stellar system, sending a rain of comets toward the inner planets and extinguishing life. Another danger in a denser galaxy is the potentially devastating effect of nearby supernova explosions. Quantifying the impact of all these factors on the density of habitable stellar systems is a challenging, but not intractable problem. At present, however, it is hard to go beyond order-of-magnitude estimates.

4. A. Vilenkin, "Predictions from quantum cosmology," *Physical Review Letters*, vol. 74, p. 846 (1995).

5. Efstathiou's approach was somewhat different from mine. He assumed that we are typical only among the presently existing observers (galaxies), while my choice was to include all observers—present, past, and future. If we are truly typical, and live at the time when most observers live, the two methods should give similar results—as in fact they do. The choice of the reference class of observers among which we expect to be typical is generally an important issue. It has been discussed in detail by the philosopher Nick Bostrom.

6. There is in fact some variation in the power of type Ia supernovae, probably due to differences in the chemical composition of the white dwarfs. But this variation can be accounted for by measuring the duration of the explosion: the power depends on the duration in a well-studied way.

7. Doppler shift is the change in frequency of electromagnetic waves when the source of waves and the observer move relative to one another. If you move toward a source of light, the frequency of the waves increases, just as a boat hits the waves more frequently as it goes against oncoming waves. The same effect occurs when the source of light moves toward a stationary observer: only the relative motion of the observer and the source is important. Quite similarly, the frequency of light emitted by a galaxy gets lower (shifts toward the red end of the spectrum) if the galaxy moves away from the observer.

8. As quoted in R. Kirshner, *The Extravagant Universe* (Princeton University Press, Princeton [N.J.], 2002, p. 221).

9. The possibility that a cosmological constant could resolve the age discrepancy between the oldest stars and the universe was advocated in the 1980s by Gerard de Vaucouleurs. More recently, it was emphasized, together with other potential benefits of a cosmological constant, by Lawrence Krauss and Michael Turner in their paper "The cosmological constant is back," published in *General Relativity and Gravitation*, vol. 27, p. 1137 (1995).

10. For a popular review of the quintessence idea, see *Quintessence: The Mystery of the Missing Mass* by Lawrence Krauss (Basic Books, New York, 2000).

11. Another problem with the quintessence model is that the flat plateau at the bottom of the hill is assumed to be at zero energy density. This amounts to the assumption that the energies of the fluctuating fermions and bosons miraculously cancel one another (see Chapter 12).

12. It is probably not an accident that we live in the disc of a giant galaxy. Galaxy formation is a hierarchical process, with smaller and denser objects merging to form larger and more dilute ones. Early dense galaxies are less suitable for life, for the reasons indicated in note 3 above.

13. This explanation of the coincidence was given in the paper I wrote with Jaume Garriga and Mario Livio, "The cosmological constant and the time of its dominance," published in the *Physical Review*, vol. D61, p. 023503 (2000). The same idea was independently suggested by Sidney Bludman, in *Nuclear Physics*, vol. A663, p. 865 (2000).

15. A THEORY OF EVERYTHING

1. Quoted in Nigel Calder, *The Key to the Universe* (Penguin Books, New York, 1977), p. 69.

2. During the 1970s and 1980s physicists tried to achieve a more unified description of particles and their interactions in the framework of the grand unified theories. The first model of this type was proposed by Howard Georgi and Sheldon Glashow of Harvard, who showed that the entire standard model, with its separate symmetries for strong and electroweak interactions, could be elegantly incorporated into a theory that had a single, but larger, symmetry pattern. Moreover, the model gave a unified description for the three basic interactions. Grand unification is a very attractive idea, and most physicists believe that it will survive as part of the final theory. But grand unified theories still have most of the shortcomings of the standard model. In particular, they require an even larger number of adjustable parameters, and gravity is still left out.

3. A broad range of issues surrounding the existence (or not) of a final theory of nature is discussed in *Dreams of a Final Theory* by Steven Weinberg (Vintage, New York, 1994).

4. An interesting possibility of an observational test of string theory comes from cosmology. Strings of astronomical size could be formed as a result of high-energy processes at the end of inflation. Like "ordinary" cosmic strings (see Chapter 6), these fundamental strings would then be accessible to observation. Strings do not emit light, so they cannot be seen directly, but they can betray their presence through their gravitational effects. Light rays from a distant galaxy located behind a long string are bent by the string gravity, and we can see two images of the galaxy next to one another, from the rays passing on the two sides of the string. Oscillating loops of string are powerful sources of gravitational waves. Existing and future gravitational wave detectors will search for their characteristic signal.

5. Recent work by Nima Arkani-Hamed of Harvard, Gia Dvali of New York University, and Savas Dimopoulos of Stanford suggests that the compact dimensions may be much larger than previously thought. In this case, the sizes of vibrating string loops are also greatly increased. The next generation of particle accelerators could then be powerful enough to reveal the "stringy" nature of the particles.

6. An eloquent expression of this philosophy, together with details of string theory, can be found in Brian Greene's book *The Elegant Universe* (Vintage Books, New York, 2000).

7. In the presence of branes, strings can have the form of closed loops, as before, but can also be open, with their ends attached to the branes. Such open string segments can move along the branes, but can never leave them. Branes play a central role in *braneworld* cosmological models, which assume that we live on a three-dimensional brane floating in a higher-dimensional space. The familiar particles, like electrons and quarks, are then represented by open strings with their ends attached to our brane.

8. The spacetime structure of expanding bubbles is similar to that of island universes, as described in Chapter 10. The bubbles are finite as viewed from the outside, but from the inside each bubble appears to be a self-contained, infinite universe. Eternal inflation with bubble island universes was envisaged by Richard Gott in 1982 and was discussed by Paul Steinhardt in a more realistic model in 1983.

9. Quoted by Davide Castelvecchi, "The growth of inflation," *Free Republic*, December 2004.
10. Leonard Susskind, interviewed by John Brockman, *Edge*, 2003.
11. Ibid.

16. DID THE UNIVERSE HAVE A BEGINNING?

1. Interesting parallels between ancient myths and scientific cosmology are discussed in *The Dancing Universe: From Creation Myths to the Big Bang* by Marcelo Gleiser (Dutton, New York, 1997).
2. A. Jinasena, *Mahapurana*, in A. T. Embree, ed., *Sources of Indian Tradition* (Columbia University Press, New York, 1988).
3. The same criticism applies to the idea of the universe coming out of chaos, as in models of chaotic inflation. This point is highlighted in the "joke" related by Timothy Ferris in his book *The Whole Shebang* (Simon & Schuster, New York, 1997). An atheist claims that the world came out of chaos, to which a believer replies, "Ah, but who made the chaos?"
4. A. K. Coomaraswamy, *Dance of Shiva* (Farrar, Straus and Giroux, New York, 1957).
5. To implement this scenario, Steinhardt and Turok introduced a scalar field with a judiciously designed energy landscape. Cosmologists are generally skeptical about their model, because the landscape appears rather contrived. Moreover, the value of the vacuum energy density, which plays a crucial role in this model, is simply put in by hand, without an explanation of why it is so small or why it dominates the universe at about the epoch of galaxy formation.
6. This method of proving spacetime incompleteness by showing that certain past- or future-directed histories have a finite duration dates back to Hawking and Penrose's work in the 1960s and '70s.
7. One way to avoid the conclusion of the theorem is to assume that the expansion rate gets smaller and smaller as we go backward in time, so the universe becomes static at past infinity. This sort of scenario was suggested in 2004 by George Ellis and his collaborators. They assumed that the universe started out as a static Einstein world. The problem, however, is that Einstein's universe is unstable and could not have existed for an infinite time. (See note 3 to Chapter 2 on p. 209.)
8. Another interesting attempt to avoid the beginning of the universe was made in the 1998 paper "Can the universe create itself?" written by J. Richard Gott and Li-Xin Li of Princeton University. (The paper is published in *Physical Review D*, vol. 58, p. 023501.) Gott and Li suggest that as one goes backward in time, one gets caught in a time loop, going through the same events over and over again. Einstein's general relativity does allow, in principle, the existence of closed loops in time. (For an entertaining discussion, see Richard Gott's wonderful book *Time Travel in Einstein's Universe*.) However, as Gott and Li themselves point out, in addition to histories circling in a loop, the spacetime they envisage necessarily contains some incomplete histories, like the space traveler's history discussed in the text. This means that the spacetime itself is past-incomplete, and therefore does not provide a satisfactory model of a universe without a beginning.

9. A. Borde, A. H. Guth, and A. Vilenkin, "Inflationary spacetimes are not past-complete," *Physical Review Letters*, vol. 90, p. 151301 (2003).

10. E. A. Milne, *Modern Cosmology and the Christian Idea of God* (Clarendon, Oxford, 1952).

11. Pope Pius XII, Address to the Pontifical Academy of Sciences, November 1951; English translation is in P. J. McLaughlin, *The Church and Modern Science* (Philosophical Library, New York, 1957). The pope's enthusiasm was not universally shared by all clergy. In particular, Georges Lemaître, who was both a Catholic priest and a renowned cosmologist, thought that religion should keep to the spiritual world, leaving the material world for science. Lemaître even tried to talk the pope out of endorsing the big bang. It appears that in later years the pope had second thoughts about his remarks. Neither he nor his successors ever repeated this attempt at direct verification of religion by science.

12. As quoted in C. F. von Weizsacker, *The Relevance of Science* (Harper and Row, New York, 1964).

17. CREATION OF UNIVERSES FROM NOTHING

1. A. Vilenkin, "Creation of universes from nothing," *Physics Letters*, vol. 117B, p. 25 (1982). I later learned that the possibility of spontaneous nucleation of the universe from nothing was discussed about a year earlier by Leonid Grishchuk and Yakov Zel'dovich of Moscow State University in Russia. However, they did not offer any mathematical description for the nucleation process.

2. This story is based on a conversation I had with Edward Tryon when I visited him in New York in October of 1985.

3. At about the same time, an idea very similar to Tryon's was put forward by Piotr Fomin of the Institute for Theoretical Physics in Kiev, Ukraine. In fact, the sequence of steps shown in Figure 17.3 was not clearly spelled out by Tryon and first appeared in Fomin's paper. Unfortunately, Fomin had trouble finding a journal that would publish his work. In the end it was published in 1975 in an obscure Ukrainian physics journal.

4. E. P. Tryon, "Is the universe a vacuum fluctuation?" *Nature*, vol. 246, p. 396 (1973).

5. In the late 1970s and early '80s there were some attempts to develop mathematical models of quantum creation from the vacuum. Richard Brout, François Englert, and Edgard Gunzig of the Free University of Brussels suggested in 1978 that superheavy particles, 10^{20} times heavier than the proton, could be spontaneously created in the vacuum. The particles would curve space, the growing curvature would trigger further particle creation, and the process will extend to a larger and larger region as an expanding bubble. Inside the bubble, the heavy particles will quickly decay into light particles and radiation, resulting in an expanding universe filled with matter. This model has the same problem as Tryon's scenario: it does not really explain the origin of the universe. If flat empty space were indeed so unstable, it would be rapidly filled with expanding bubbles. Such an unstable space could not have existed forever and cannot, therefore, be taken as the starting point of creation.

David Atkatz and Heinz Pagels of Rockefeller University wrote a paper in 1982, suggesting that before the big bang the universe existed in the form of a small spherical space packed with exotic high-energy matter—a sort of "cosmic egg." They designed a model in which the "egg" was classically stable, but could tunnel to a bigger radius and expand. (To my knowledge, this was the first discussion of quantum tunneling of the universe as a whole.) Once again, the problem is that the unstable "egg" could not have existed forever, and we are left with the problem of where the egg came from.

6. A. H. Guth, *The Inflationary Universe* (Addison-Wesley, Reading [Mass.], 1997, p. 273).

7. Saint Augustine, *Confessions* (Sheed and Ward, New York, 1948).

8. A. Vilenkin, "Quantum origin of the universe," *Nuclear Physics*, vol. B252, p. 141 (1985).

9. I am grateful to Ernan McMullin for emphasizing to me the importance of requiring that the universes in the ensemble must be really existing, not merely possible universes.

10. J. B. Hartle and S. W. Hawking, "The wave function of the universe," *Physical Review*, vol. D28, p. 2960 (1983). Hawking outlined the basic idea of this work about a year earlier, in *Astrophysical Cosmology: Proceedings of the Study Week on Cosmology and Fundamental Physics*, edited by H. A. Bruck, G. V. Coyne, and M. S. Longair (Pontifica Academia, Vatican, 1982), but at that time he did not provide any mathematical details.

11. A firsthand account of the no-boundary proposal can be found in Hawking's bestselling book *A Brief History of Time* (Bantam, New York, 1988, p. 136).

12. One caveat is that the string theory landscape may consist of several disconnected domains, with no possibility for bubbles from one domain to nucleate in another. Then bubbles formed during eternal inflation will only contain vacua belonging to the same domain as the initial vacuum that filled the universe when it came into being. In this case, the nature of the multiverse does depend on the initial state, and a test of quantum cosmology is in principle possible.

18. THE END OF THE WORLD

1. Physical processes in the distant future of the universe have been studied by Martin Rees and Don Page, among others. For a popular review, see the book by Paul Davies, *The Last Three Minutes: Conjectures about the Ultimate Fate of the Universe* (Basic Books, New York, 1994).

2. This scenario is based on the analysis by K. Nagamine and A. Loeb in "Future evolution of nearby large-scale structure in a universe dominated by a cosmological constant," *New Astronomy*, vol. 8, p. 439 (2003).

3. The prediction that the local region of the universe will collapse to a big crunch was made in the paper I wrote with Jaume Garriga, "Testable anthropic predictions for dark energy," *Physical Review*, vol. D67, p. 043503 (2003). We pointed out, however, that this prediction was not likely to be tested anytime soon.

19. FIRE IN THE EQUATIONS

1. Alan L. Mackay, *A Dictionary of Scientific Quotations*, Institute of Physics Publishing, Bristol [U.K.], 1991.

2. This situation, that an infinite ensemble is much simpler than one of its members, is very common in mathematics. Consider, for example, the set of all integers: 1,2,3, ... It can be generated by a simple computer program, which takes only a few lines of code. On the other hand, the number of bits needed to specify a specific large integer is equal to the number of digits required to write it in a binary code, and can be much larger.

3. P.A.M. Dirac, "The evolution of the physicist's picture of nature," *Scientific American*, May 1963.

4. For an interesting discussion of beauty in scientific theories, see *The Accelerating Universe: Infinite Expansion, the Cosmological Constant, and the Beauty of the Cosmos* by Mario Livio (Wiley, New York, 2000).

5. Needless to say, "simplicity" and "depth" are almost as difficult to define as "beauty."

6. M. Tegmark, "Parallel universes," *Scientific American*, May 2003.

7. Tegmark makes no distinction between mathematical structures and the universes they describe. He argues that mathematical equations describe every aspect of the physical world, so that each physical object corresponds to some entity in the Platonic world of mathematical structures and vice versa. In this sense the two worlds are equivalent to one another, and Tegmark's view is that our universe *is* a mathematical structure.

8. To address this problem, Tegmark suggested that mathematical structures might not all be equal; they might be assigned different "weights." If these weights rapidly decline with increasing complexity, the most probable structures might be the simplest ones that can still contain observers. The introduction of weights may resolve the complexity problem, but then we are faced with the question, Who determines the weights? Should we recall the Creator from his exile? Or should we perhaps enlarge the ensemble of mathematical structures still further, to include all possible weight assignments? I am not sure that the notion of weights for the set of all mathematical structures is even logically consistent: it seems to introduce an additional mathematical structure, but all of them are supposed to be already included in the set.

9. Depending on the fundamental theory, the constants may vary within the island universes as well. Our own island universe is then mostly barren, with rare habitable enclaves.

Acknowledgments

My friends and colleagues, whose opinion is very important to me, read the manuscript and kindly offered their critique and suggestions. Alan Guth, Steven Weinberg, and Jaume Garriga gave me their advice and very useful comments about parts of the book. Paul Shellard and Ken Olum provided extensive feedback on the entire text, and straightened me out on some important details of science. I am deeply grateful to all of them.

Special thanks to Delia Schwartz-Perlov, who turned my sketches into wonderful illustrations, refined some of my cartoons, and suggested many improvements in the text. I also benefited from stimulating correspondence with Frank McCormick and Max Tegmark.

Thanks to my editor, Joseph Wisnovsky, for his enthusiasm for the project and guidance throughout the production of this book. Many thanks to Vitaly Vanchurin, who was always ready to help whenever I ran into trouble with my computer, to Marco Cavaglia and Xavier Siemens for historical references, and to Susan Mader for her assistance with photographs. I also owe a debt of gratitude to Susan Rabiner for her vital advice at the early stages of this work.

Closer to home, my thanks go to Joshua Knobe and my daughter, Alina, for their useful suggestions, enthusiasm, and support, and to my wife, Inna, who served as editor, critic, and trusted advisor.

Index

acceleration, cosmic, 148–49, 212n1

Albrecht, Andreas, 61, 63

alchemy, 34–35

Alfonso the Wise, King of Castile, 199

Alpha Centauri, 94, 95

Alpher, Ralph, 35, 39, 40

Andromeda galaxy, 77, 197

anthropic selection, 127, 132–35, 149,
  150–52, 210n5
  cosmological constant and, 135–39,
    145–46, 162
  predictions drawn from, 140–44,
    151
  string theory and, 162–65

antigravity, 73–76

antiparticles, 33, 34

antiquarks, 154

Archimedes, 201

Aristotle, 15

Arkani-Hamed, Nima, 218n5

Aryal, Mukunda, 84, 212n3

Astrophysical Journal, 216n3

Atkatz, David, 220n5

atomic nuclei, 30, 48, 49, 130, 153, 211n3
  formation of, 35–37
  impact of heat on, 32

atoms, 34, 115, 153, 211n3
  formation of, 38
  see also atomic nuclei

Augustine, Saint, 186, 221n7

Babson, Roger W., 73–76

bacteria, reproduction of, 80,
  81

Barcelona, University of, 102

Bardeen, Jim, 63

Barrow, John, 215n2

Bekenstein, Jacob, 108

bell curves, 141–42, 151

Bell Telephone Laboratories, 39

Bethe, Hans, 35

big bang, 4–5, 9–11, 16, 28, 43, 66, 111,
  170, 195, 204, 220n5
  afterglow of, 90
  cosmic radiation and, 39–41
  elements formed in, 35–37, 131
  evidence of, 10
  and Friedman's solutions to Einstein's
    equations, 22, 24, 27
  as hot, 33–38
  inflationary universe and, 46, 47, 52,
    82, 93–99, 106, 117

big bang (continued)
  nature of, 11–12, 47–48
  repulsive gravity and, 11–12
  theological response to, 176–77
"big crunch," 22, 27, 173, 194–98, 204,
    221n3
biology, 40
black holes, 68, 92, 108, 112, 195, 210n4
  evaporation of, 195, 196
Bludman, Sidney, 217n13
Bohr, Niels, 5, 30, 31, 115
Boltzmann, Ludwig, 25–26, 210n5
Bondi, Herman, 27–28
Borde, Arvind, 174–76, 220n9
Bose, Satyendra, 124n
bosons, 124, 137, 158, 214n2, 217n11
Bostrom, Nick, 135, 215n6, 217n5
Bousso, Raphael, 162, 164
branes, 161, 162n, 181, 219n7
British Columbia, University of, 193n
Brout, Robert, 67n, 220n5
Brundrit, G., 114, 213n3
Brussels, Free University of, 220n5
bubble nucleation, 162–65, 181–82,
    192–93, 198, 212n2, 218n8, 220n5,
    221n12
Bush, George W., 113n

Caesar, Julius, 112
California, University of
  Berkeley, 162
  Santa Barbara, 162, 188
California Institute of Technology
    (Caltech), 154, 158
Cambridge University, 27, 61–65, 133,
    171, 192, 210n6
carbon, 34, 37, 131
Carr, Bernard, 215n2
Carter, Brandon, 133, 134, 135n, 143,
    215n3
Casimir, Hendrik, 214n1
CERN (European Center for Nuclear
    Research), 154

Chandrasekhar limit, 147
chaotic inflation, 84–86, 219n3
charge, conservation of, 33
Chibisov, Gennady, 65, 212n3
Chicago, University of, 158
Churchill, Winston, 165
civilizations, 134n
  future of, 93–98
  identical, 117
Clinton, Bill, 116, 213n7
closed-universe model, 20, 22,
    184–85
Clover Observatory, 91, 213n4
coarse-grained description, 107–108,
    110–11
COBE satellite, see Cosmic Background
    Explorer (COBE) satellite
coincidence problem, 125–27, 149–50
Coleman, Sidney, 56, 83, 212n2
Communist Party, French, 31
compactification, 159–60
complex numbers, 202n
Conference on General Relativity and
    Gravitation (Padova, 1983), 188
Confessions (Augustine), 186
conservation laws, 33, 112, 125
constants of nature, 128–32, 151, 165,
    215n2
  anthropic selection and, 132–35, 142,
    152
  eternal inflation and, 144
  see also cosmological constant
Copenhagen interpretation, 115, 116n,
    187
Copernicus, Nicolaus, 117
Cosmic Background Explorer (COBE)
    satellite, 40, 42, 90
cosmic background radiation, 10, 39–42,
    46, 88–91, 195
  see also microwaves, cosmic
cosmic egg, 169–70, 221n5
cosmic horizon, 108
Cosmo-98 conference (Monterey), 192

cosmological constant, 24n, 87–90, 124–27, 129, 144–49, 161, 164, 197, 214n2, 217n9
accelerated expansion and, 148–49
anthropic selection and, 135–39
Einstein and, 19, 20, 22, 88, 124, 194
principle of mediocrity and, 145–46, 150, 162
Coulomb, Charles-Augustin de, 155
creation stories, 169–70
Curie, Marie, 31
curvature of spacetime, 16–18, 23
cyclic universe, 171–72, 175

Dalí, Salvador, 104–105, 113
Damour, Thibault, 142
dark energy, 88n
dark matter, 43, 88
Darwin, Charles, 67
Davies, Paul C. W., 215n2, 221n1
de Sitter, Willem, 173, 210n2
de Sitter spacetime, 172–73
density perturbations, 77, 90, 106, 213n3
calculation of, 62–66
parameter Q of, 145, 216n3
see also inhomogeneities, cosmic
deuterium, 35–36, 131
Deutsch, David, 213n4
Dicke, Robert, 39–40, 133, 215n5
dimensions, extra, 159–60
Dimopoulos, Savas, 218n5
Dirac, Paul, 150–51, 153, 201, 215n5, 222n3
disorder, 25–26
distance determination, 147
DNA, 128
doomsday argument, 143, 216n1
Doppler shift, 88n, 147, 217n7
double-slit experiment, 109–10
Dvali, Gia, 218n5
dwarf galaxies, 197, 200, 216n3

$E=mc^2$, 11–12, 14, 49
Earth
evolution of intelligent life on, 132
human population of, 204
infinite duplicates of, 112–14, 117
magnetic field of, 122
École Normale Supérieure, 158
economic inflation, 51–52, 212n2
Edison, Thomas, 74
Efstathiou, George, 146, 217n5
Ehrenfest, Paul, 17, 55
Einstein, Albert, 13–20, 50, 68, 135, 156, 209n3, 210nn2, 3, 211n2, 219nn7, 8
cosmological constant of, 19, 20, 22, 88, 124, 194
Friedman's solutions to equations of, 21–22, 27, 172, 182, 210n3
mass-energy relation ($E=mc^2$) of, 11–12, 14, 49
on observer dependency of time order of events, 94, 95, 175
Stalinist rejection of theories of, 30
see also relativity theory
electric force, 155
electromagnetism, 38, 49, 128, 129, 132n, 154–55, 215n5, 217n7
in final theory of nature, 152
quantum fluctuations of, 121–24, 214n1
in supernovae, 131
electrons, 33–35, 38, 43, 49, 88, 124, 130, 153, 211n3, 214n2
annihilation into photons of, 195
magnetic moment of, 150–51
mass of, 128, 145
in string theory, 156
virtual, 184
electroweak force, 49, 52, 61, 67, 69, 154–55
elementary particles, 48–49
collisions of, 33
masses of, 128

elements
   chemical properties of, 152–53
   origins of, 34–38, 131
   periodic table of, 153–55
elliptical galaxies, 200
Ellis, George, 114, 213n3, 219n7
empty space, gravity of, 19–20
energy conservation, law of, 11–12, 33,
   180
Englert, François, 67n, 220n5
entropy, 25, 26, 171, 172, 210n4
equilibrium, thermal, 25, 171
eschatology, cosmic, 194
eternal inflation, 80–84, 91, 100, 101,
   117, 172, 200, 212n4, 214n8
   computer simulation of, 84, 85
   and future of intelligent life, 93–97
   island universes and, 114, 170,
      203–204, 218n8
   parallel universes and, 116
   necessity of a beginning, 173, 175
   quantum processes during, 137–38,
      144
   string theory and, 162, 192–93, 221n12
   time and, 97–99
   versus end of universe, 196
Euclidean geometry, 24, 87, 202
Euclidean time, 182, 190–91
Euler's formula, 201–202
European Center for Nuclear Research
   (CERN), 154
Everett, Hugh, III, 115
Everett interpretation of quantum
   mechanics, 115–16, 187
evolution, Darwin's theory of, 67

false vacuum, 48–53, 84, 85, 121, 170,
   173, 175, 178, 212n2, 213n4
   in "chaotic" inflation, 85
   decay of, 56–58, 79–82, 97, 99
   energy density of, 69, 91, 172, 179,
      181, 214n2
   island universes in, 100, 101

   in parallel universes, 187
   repulsive gravity of, 48, 50, 68, 76, 179
   strings of, 64
   tunneling and, 186
Fermi, Enrico, 124n, 154
fermions, 124, 137, 214n2, 217n11
Ferris, Timothy, 219n3
final theory of nature, 135, 152–65,
   211n2
   gravity and, 155–56
   mathematics of, 202–203
   search for, 152–55
   see also string theory
fireball, cosmic, 29, 31–41, 170
   in cyclic universe, 171
   at end of universe, 195
   ignited by decaying false vacuum, 82
   images of, 90
five-nucleon gap, 36, 37
flatness problem, 52, 53, 63, 87, 89
fluxes, 161n
foam, spacetime, 123, 156, 158, 184
Fomin, Piotr, 220n3
Ford, Larry, 75
fractals, 84
French Communist Party, 31
Friedmann, Alexander, 21–22, 27, 34, 66,
   97–98, 178–80, 182, 210nn2, 3
   big crunch predicted by models of,
      194
   Gamow and, 29
   on geometry and density of universe,
      22–24, 47, 88–89, 197
fundamental theory of nature, see final
   theory

galaxies, 131, 200, 209n3, 216n3
   clusters of, 196, 197
   in contracting universe, 195
   dark matter and, 88
   distant, 5, 10, 28, 42
   distribution of, 26, 77, 106
   duplicate, 112

formation of, 4, 10, 41–43, 62, 106, 138–39, 145–46, 150, 171, 196
helium content of, 36
in island universes, 99
in open universe, 23
properties of, 128
radio emissions from, 210n6
type Ia supernova explosions in, 147
Galileo, 15, 17
gamma rays, 38
Gamow, George, 16, 29–31, 34, 35, 37, 39, 40, 58, 66, 210n2
Gamow, Lyuba, 30
Garriga, Jaume, 102–105, 111, 113, 213n1, 217n13, 221n3
gases, expansion of, 31–32
gauge symmetry, 125
Gell-Mann, Murray, 111, 154
general theory of relativity, see relativity theory, general
genetic code, 40
geometry, 21–24, 47, 89, 91, 96, 201
    Euclidean, 24, 87, 202
    spacetime, 184
George Washington University, 31
Georgi, Howard, 218n2
giant galaxies, 200, 216n3
Glashow, Sheldon, 154, 218n2
Gleiser, Marcelo, 219n1
Gliner, Erast, 67n
gluons, 155, 158
God
    and cosmic egg scenario, 169–70
    before creation, 186
    existence of, 176–77
    as mathematician, 200–202
gold, 34, 35
Gold, Thomas, 27–28
google, 52
googleplex, 111, 112
Gore, Al, 113n
Gott, Richard, 143, 216n1, 218n8, 219n8
Göttingen, University of, 29–30

graceful exit problem, 54–56, 60–62, 67
grand unified theories, 218n2
grand-unified vacuum, 49, 52, 61, 69
Grassmann, Hermann, 124n
Grassmann numbers, 124
gravitational constant, 145
gravitational waves, 91, 96, 213nn3, 4
gravitons, 155–56, 158
gravity, 15–19, 27, 47, 52–53, 73–76, 108, 128, 154, 175, 209n3, 211n3, 215n5
    amplification of perturbations by, 106
    of dark matter, 42, 88
    instability of, 41
    Newton's theory of, 202
    quantum, 155–56, 183
    and rate of cosmic expansion, 148
    repulsive, see repulsive gravity
    strength of, 131–32
    in string theory, 158
Great Depression, 74
Greeks, ancient, 200–201
Green, Brian, 218n6
Green, Michael, 158
Griffiths, Robert, 111
Grishchuk, Leonid, 220n1
Gross, David, 165, 216n8
Guinness Book of Records, 140
Gunzig, Edgard, 67n, 220n5
Guth, Alan, 9–12, 51–53, 73, 81n, 82–83, 162–63, 174–76, 185, 209n1, 211nn1, 4, 212nn2, 3, 213n3, 220n9, 221n6
    calculation of density perturbations by, 62–65
    graceful exit problem recognized by, 55–56
    on repulsive gravity, 11–12, 48, 50
    success of inflation theory of, 66–67, 87

half-life, 80–81
Hartle, James, 111, 188–90, 221n10
Harvard University, 9, 52, 56, 73, 154, 191, 212n1, 218nn2, 5

Hawking, Stephen, 27, 61–65, 188–93, 195, 219n6, 221nn10, 11
heat-death problem, 32, 171, 172, 175
Heisenberg, Werner, 107
Heisenberg's uncertainty principle, 107, 184
helium, 34–38, 82, 131, 210n4
Helmholtz, Hermann von, 25, 210n4
Herman, Robert, 35, 39, 40
High-Redshift Supernova Search Team, 148, 212n1
Hinduism, 169, 171
histories, 108–13
Hogan, Craig, 128, 215n1
horizon, cosmic, *see* cosmic horizon
horizon problem, 46, 52, 53, 54, 63
horoscopes, 199
Hoyle, Fred, 27–28, 37
Hubble, Edwin, 10, 24
human race, longevity of, 143
Hunter College, 183
hydrogen, 34–37, 131, 139, 210n4
   heavy, 35

imaginary numbers, 182, 202
inertial motion, 15
inflation, 9, 45–53, 67–69, 121, 136, 170, 178–79, 188, 203, 212n2
   chaotic, 84–86, 219n3
   density perturbations from, 62–66, 77–80, 90, 106
   eternal, *see* eternal inflation
   graceful exit problem in, 54–56, 60–62, 67
   no-boundary proposal and, 191, 192
   observational evidence for, 87–92, 116–17
   quantum fluctuations during, 108, 213n3
   repulsive gravity and, 11–12, 48, 50
   success of theory of, 66–67
   tunneling and, 179–83, 186, 187, 191, 192

inflaton, 136
inhomogeneities, cosmic, 41, *see also* density perturbations
instability, gravitational, 41
Institut des Hautes Études Scientifiques, 142
intelligent life, 145, 199
   evolution of, 106, 132
   future of, 93–98
interference, quantum, 110
International Congress on Nuclear Physics (Rome, 1930), 30
inverse square law, 17–18
iron, 36, 131
island universes, 81–82, 170, 203–204, 218n8, 222n9
   computer simulation of, 84, 85
   in de Sitter spacetime, 173
   frontiers of, 93–97
   time in, 98–100

Jainism, 169
Jinasena, 169–70, 177

Kac, Victor, 201
kelvin scale, 32n
Kepler, Johannes, 26, 201
Khvolson, Orest, 29
"kickspan," 79
Kiev Institute for Theoretical Physics, 220n3
kicks, quantum, 62, 65–66, 77–80, 85–86, 137
Kirshner, Robert, 212n1
Knobe, Joshua, 214n9
Kragh, Helge, 210n3
Krauss, Lawrence, 217nn9, 10
Krutkov, Yuri, 210n3, 212n3

Landau, Lev, 63, 142, 212n3
Langevin, Paul, 31
last scattering, 42, 126
Lebedev Institute, 65

Lemaître, Georges, 19*n*, 24*n*, 220*n11*
Leslie, John, 143, 215*n1*, 216*n1*
Li, Li-Xin, 219*n8*
life, evolution of, 106, 131, 204
    anthropic selection and, 133–34, 138
    in closed universe, 185
    constants of nature and, 144–45
light
    from distant galaxies, 42, 52
    Doppler shift of, 88*n*, 147
    quantum nature of, 107
    spectrum of, 36, 38–39
    speed of, 14, 38, 46, 94–96, 175
light-years, 41
Linde, Andrei, 58–61, 63, 67*n*, 84–86,
    94, 136, 137, 164, 192, 198, 212*n4*,
    213*n1*
lithium, 36, 131
Livio, Mario, 217*n13*, 222*n4*
Local Group of galaxies, 197
Loeb, A., 221*n2*
logarithms, "natural," 202
Long Island University, 174
Lou Gehrig's disease, 63
Ludwig-Maximilians University, 60,
    65*n*

magnetic fields, *see* electromagnetism
many-worlds theory, 114–16
    see also multiverse hypothesis; parallel
    universes
Martel, Hugo, 146
Marxism-Leninism, 30
Massachusetts Institute of Technology
    (MIT), 51, 53, 82, 202
mathematics, nature and, 200–203
Mayer, Jean, 75
McCarthy, Kathryn, 186
McMullin, Ernan, 221*n9*
mediocrity, principle of, 143–44, 146,
    149, 150, 151, 164, 165, 203
Mendeleyev, Dmitry, 153, 155
Mermin, David, 213*n6*

messenger particles, 155, 158
microwaves, cosmic, 4, 10, 38–43, 45, 66,
    88–89
    polarization of, 91
Milky Way galaxy, 77, 197
    replicas of, 112
Milne, Edward, 177, 220*n10*
Minkowski, Hermann, 15–16
Mohammed, Gul, 140
Moon, 90, 132
Moscow State University, 220*n1*
motion
    inertial, 15–16
    planetary, 17, 201
Mukhanov, Slava, 60, 65, 148, 212*n3*
multiverse hypothesis, 133, 134, 144,
    151, 203–204, 221*n12*
muons, 34, 154

Nagamine, K., 221*n2*
Nambu, Yoichiro, 158
nanosecond, 122
Nature, 143–44, 184
Ne'eman, Yuval, 154
Nernst, Walter, 177
neutrinos, 49, 195
    mass of, 145
    weakly interacting, 131
neutrons, 32, 35–36, 49, 87–88, 131*n*,
    153, 211*n3*
    mass of, 128, 130
neutron stars, 195, 210*n4*
New Scientist, 113
Newton, Isaac, 17, 74, 106, 145, 155,
    202
New York University, 218*n5*
Niels Bohr Institute, 158
Nielsen, Holger, 158
night sky paradox, 26
Nobel Prize, 35, 40, 41, 63, 113, 135,
    154, 165, 177
nuclear physics, 30, 31
nuclear reactions, 35–37, 115, 132

nucleation, 220n1
  see also bubble nucleation
nucleons, 32, 34, 35, 38, 43
  decay of, 196
Nuffield Workshop (Cambridge, 1982),
  61–65, 212n3

Olum, Ken, 75, 214n9
Omega parameter, 195–96
Omnes, Roland, 111
O-regions, 106–14, 116, 204
  histories of, 108–13
  possible states of, 107–108
Oxford University, 146
oxygen, 131

Page, Don, 221n1
Pagels, Heinz, 220n5
pair annihilation, 33, 34
parallel universes, 114–16, 187–88, 199
  mathematical structures of, 202–203
particle physics, 33–34, 48–49, 115, 137,
  139, 199, 214n2
  calculation of magnetic moment in,
    151
  cosmic strings in, 64
  cosmological constant and, 88, 124, 126
  gauge symmetry in, 125
  scalar fields in, 56
  search for final theory in, 152–55
  Standard Model of, 129, 155
  string theory and, 160, 151
  strong and electroweak interactions
    in, 67
  variable constants in, 144
  see also elementary particles
pendulum, dynamics of, 29
Penrose, Roger, 27, 219n6
Penzias, Arno, 39–41
periodic table, 153–55
Perlmutter, Saul, 148, 212n1
Perry, Malcolm, 183
Petrograd University, 21

photons, 32–34, 38, 107, 110, 124, 214n2
  interaction of, 131, 154, 155
Physical Review, 83
Physical Review Letters, 176
Physics Letters, 84
pi, 202
Pi, So-Young, 62
Pius XII, Pope, 177, 220n11
Planck, Max, 40, 91n
Planck length, 123, 156, 158, 159, 184,
  185, 187
Planck satellite, 91
planets
  formation of, 131, 133
  motion of, 16, 201
Plato, 15
pocket universes, *see* island universes
polarization, 91
Polchinski, Joseph, 146, 162, 164
Pontifical Academy of Sciences, 177
Popper, Karl, 134
Port Alguer (Cadaqués) (Dalí), 105
positrons, 33, 34, 124, 154
  annihilation into photons of, 195
  virtual, 184
primeval fireball, *see* fireball, cosmic
Princeton University, 39–40, 42n, 143,
  146, 182–83, 219n8
probabilities, quantum-mechanical, 115,
  187–89
protons, 32, 35–36, 49, 87, 131n, 153,
  154, 211n3, 215n5
  in hydrogen nucleus, 35
  mass of, 128, 130
Ptolemy, 199
Pythagoras, 200–201
Pythagorean theorem, 213n2

quantum kicks, 62, 65–66, 77–80, 85–86,
  137
quantum theory, 29–30, 33, 38, 62, 66,
  91n, 152–54, 201, 209n3
  gravity in, 155–56, 158

histories in, 109–12
parallel universes in, 114–16
probabilities in, 115, 187–89
tunneling in, *see* tunneling, quantum
uncertainty in, 106–107
vacuum in, 121
quarks, 32, 34, 49, 124, 154, 155,
214n2
masses of, 131n
in string theory, 156
quarternion, 202
Queen Mary College, 158
QUIET Observatory, 91
quintessence model, 149, 217n11

Rabi, Isidor, 115
radiation, 35, 195
cosmic, 10, 38–43, 88–89
electromagnetic, 38, 121–23
radioactivity, 30
radio astronomy, 39
radio waves, 28, 38
redshift, 88n, 147
Rees, Martin, 38, 133, 134, 211n6, 215n2,
216n3, 221n1
relativity theory, 30, 94, 175
general, 15–19, 21, 27, 29, 68, 92, 156,
172, 219n8
special, 13–14
repulsive gravity, 11–12, 19–20, 27, 88,
197, 209n3
anthropic selection and, 138
in de Sitter spacetime, 172
of false vacuum, 48, 50, 68, 76, 179
Riess, Adam, 148
Rockefeller University, 220n5
Roman Catholic Church, 177,
220n11
Rosenfeld, Leon, 29
Rubakov, Valery, 192n
Russian Revolution, 21
Rutherford, Ernest, 153
Ryle, Martin, 210n6

Sakharov, Andrei, 113, 213n2
Salam, Abdus, 154
Sato, Katsuhiko, 67n
scalar fields, 56–63, 65–68, 97, 121–22,
136, 144, 161, 164, 188, 211n2,
212n2
in chaotic state, 84–85
quintessence, 149
and end of universe, 198
energy landscape of, 117, 219n5
random walk of, 77–80, 82, 85–86, 137
Scherk, Joel, 158
Schmidt, Brian, 148, 212n1
Schwartz, John, 158
Shapiro, Paul, 146
Siding Springs Observatory, 212n1
Simpsons, The (television show), 192n
singularities, 22, 82n, 179, 182
initial, 24, 27, 47, 63
solar system, 117
formation of, 35
solids, properties of, 115
Solvay Congress (Brussels, 1933), 31
Sommerfeld, Arnold, 18
sound waves, 48–49
Space Telescope Science Institute,
148
spacetime, 91, 94–96, 98–101, 133, 182,
213n2, 219n6
curvature of, 16–18, 23, 156, 211n2
de Sitter, 172–73
eternally inflating, *see* eternal
inflation
quantum fluctuations in, 123, 156,
158, 184
singularities, 22
without past boundary, 189–92
special theory of relativity, *see* relativity
theory, special
Spinoza, Baruch, 13
spiral galaxies, 42, 150, 200
Stalinism, 30
standard candles, 147

Standard Model, 155, 218n2
Stanford University, 9, 218n5
Starobinsky, Alexei, 63, 67n, 192n, 211n2,
    213n3, 216n9, 218n8, 219n5
stars, 41, 131–32, 210n4, 215n4, 216n3
  age of, 149, 217n9
  dark matter and, 88
  death of, 93, 94, 134, 195
  element formation in, 35–37
  formation of, 94, 133
  masses of, 132
steady-state cosmology, 27–28, 37, 171
Steinhardt, Paul, 61, 63–65, 149, 165,
    171–72, 216n9, 218n8, 219n5
stock market crash of 1929, 74
strings, cosmic, 61, 64, 218n4
string theory, 156–59, 181, 193, 201, 202,
    211n2, 218n4
  landscape of, 159–65, 192–93, 198,
    221n12
  see also superstring theory
strong force, 49, 67, 128, 129, 131, 152,
    153, 155, 158
Sun, 37, 117
  lifetime of, 132–34, 215n4
  motion of planets around, 16, 201
  nuclear reactions in, 35
superconductivity, 115
Supernova Cosmology Project, 148,
    212n1
supernovae, 37, 88, 131, 133, 139,
    147–48, 217n6
superstring theory, 61, 108, 146
supersymmetric theories, 61, 214n2
surface of last scattering, 42
Susskind, Leonard, 108, 158, 164, 165,
    219n10

Tegmark, Max, 202–203, 216n3,
    222nn6, 7, 8
temperature, scalar field of, 56
tension, 50
Texas, University of, 146

Theory of Everything, see final theory
  of nature
thermal equilibrium, 25, 171
thermal fluctuations, 26
thermodynamics, second law of, 25, 171
't Hooft, Gerard, 108
time, 97–100, 180–81, 186
  Euclidean, 182, 190–91
  see also spacetime
Tipler, Frank, 215n2
top quark, 34
tritium, 36
true vacuum, 49, 50, 54, 77, 84, 85
  energy density of, 121–26, 149–50,
    197–98 (see also cosmological
    constant)
  quantum fluctuations of, 151
Tryon, Edward, 183–86, 220nn2, 3, 4
Tufts University, 73–76, 186
  Institute of Cosmology, 75–76, 82,
    176, 182
tunneling, quantum, 30, 57–58, 179–83,
    186–88, 191, 192, 204–205, 212n2,
    221n5
Turner, Michael, 63, 217n9
Turok, Neil, 171–72, 219n5
Twain, Mark, 135

uncertainty, quantum, 106–107, 184
unified theory, see final theory of nature
universe
  age of, 133–34, 149, 217n9
  beginning of, 26–27, 169–77 (see also
    big bang)
  bubbling, 162–65, 181–82, 192–93
  closed, 20, 22, 23
  cyclic, 171–72, 175
  density of, 47, 87
  disordered, 25–26
  exhaustive randomness of, 114
  expanding, 10, 12, 20, 22, 24, 28, 29,
    32, 38, 50–52, 54, 88, 148, 179 (see
    also inflation)

fine-tuning of, 129–32, 134, 135, 200, 215n2
geometry of, 21–24, 47, 87, 89, 91, 96
observable, end of, 194–98
Ptolemy's model of, 199
as quantum fluctuation, 183–86, 188
spherical, 87, 89
steady-state theory of, 27–28, 37
structure of, 18–19
super-large-scale view of, 80
without past boundary, 189–92
see also island universes, parallel universes
Unruh, Bill, 193
Upanishads, 169
uranium, 34, 35n

vacuum, 48–50, 211n2, 221n12
in cyclic universe, 171
creation of matter out of, 28
decay of, 54, 56–58
fluctuations in, 183–86, 214n1
gravity of, 19–20, 209n3
in string theory, 161–65
see also false vacuum; true vacuum
Vanhurin, Vitaly, 75, 84
Vaucouleurs, Gerard de, 217n9
virtual particles, 184
visible light, 38, 39

Wadlow, Robert Pershing, 140
weak force, 49, 128, 129, 131, 152, 154–55, 158
Weinberg, Steven, 40, 135, 138, 139, 145–46, 154, 211nn5, 7, 218n3
white dwarfs, 147, 195, 217n6
Wigner, Eugene, 201
Wilkinson, David, 42n
Wilkinson Microwave Anistropy Probe (WMAP) satellite, 42, 43, 90
Wilson, Robert, 39–41
Winitzki, Serge, 84
Witten, Edward, 165
WMAP satellite, see Wilkinson Microwave Anistropy Probe (WMAP) satellite
world lines, 16
wormholes, 68, 192
W particles, 33–34, 155, 158

X rays, 38

Yeshiva University, 158, 164
Young, Thomas, 109–10

Zel'dovich, Yakov, 192n, 220n1
Z particles, 33–34, 155, 158
Zurich Polytechnic, 13
Zweig, George, 154